MW00678866

DOGTOWN

DOGTOWN

tales of rescue, rehabilitation, and redemption

Stefan Bechtel

NATIONAL GEOGRAPHIC
WASHINGTON, D.C.

Dedication Goes here.

Library of Congress Cataloging-in-Publication

TK

ISBN: 978-1-4262-0562-0

The National Geographic Society is one of the world's largest nonprofit scientific and educational organizations. Founded in 1888 to "increase and diffuse geographic knowledge," the Society works to inspire people to care about the planet. It reaches more than 325 million people worldwide each month through its official journal, *National Geographic,* and other magazines; National Geographic Channel; television documentaries; music; radio; films; books; DVDs; maps; exhibitions; school publishing programs; interactive media; and merchandise. National Geographic has funded more than 9,000 scientific research, conservation and exploration projects and supports an education program combating geographic illiteracy.

For more information, please call 1-800-NGS LINE (647-5463) or write to the following address:

National Geographic Society
1145 17th Street N.W.
Washington, D.C. 20036-4688 U.S.A.

Visit us online at www.nationalgeographic.com

For information about special discounts for bulk purchases, please contact
National Geographic Books Special Sales: ngspecsales@ngs.org

For rights or permissions inquiries, please contact National Geographic Books
Subsidiary Rights: ngbookrights@ngs.org

Printed in the United States of America

Interior design: Cameron Zotter

09/BV6/1

Contents

Foreword by Author TK 123
Introduction by Faith Maloney 123

Chapter 1: Georgia 123
Chapter 2: Ava 123
Essay: Becoming a Dog Person, Patti Iampietro, D.V.M. 123
Chapter 3: Aristotle 123
Chapter 4: Bingo 123
Essay: *A* is for Atticus, Pat Whitacre 123
Chapter 5: Parker and Mei Mei 123
Chapter 6: Annie 123
Essay: Jenny's Gift, Sherry Woodard 123
Chapter 7: Tuffy 123
Chapter 8: Rush 123
Essay: Barnum and Sadie, Michelle Besmehn 123
Chapter 9: Bruno 123
Essay: Kaiser and Sherman, Jeff Popowich 123
Chapter 10: Knightly 123
Chapter 11: Meryl 123
Essay: Positive Reinforcement, Ann Allums 123
Chapter 12: Johnny 123
Chapter 13: Scruffy and Vivian 123
Essay: Infamous Spikey Doo, John Garcia 123
Chapter 14: Wiggles 123
Essay: A Dog for Your Lifestyle, Mike Dix, D.V.M. 123
Chapter 15: Mister Bones 123

Further Resources 123
Illustration Credits 123
Acknowledgments 123

Foreword

Sum iusci blaore tie dolortisi tie ming eugue ming estis adignis dolorpero euip et ipis do ent adiam, con ulputat ullutat nis nonsed do commy nullandre dit vulla con henibh eugiam zzriurer autat, si elessent velit ipit nisi blamet lutetum zzrit laore magna core doluptat. It aliscil iscidunt amet luptat ut non eumsan veriusto od minim irilit, quat doloreet lum zzriurem nullummy nisl dit lut praesequat wisis eugiat. Loreetu eraessisit wisl doloreetuer suscips ustrud doloreet wismolobor sequipisl ent nos ea faci euis augait, quisi.

Henis nosto doloreet velit praesto consequisi euis nos dignisim zzril doleniam dunt auguer atinibh exercilla facilit venis dionulput alit, quate mod delit, sisi tet ut ametum zzrit luptat wiscipi ssequipit am delis autpat nit praessectet nonsequ amcore consequam zzrilisl utem nit iriustrud erosto deliquisl dolestisl endre magnis eugiat. Ugait venibh exer susci esed magna ad et amet volobor illut ute te magna consed er adigna consenim doloreet wismodolutem dolore tet dolore magnis elessectem am zzrilit, quat accum ip eniam augait augait, quam zzrit eugait augait loreet in ent iusciduiscin henim doloreet dolorperat atie dolumsa ndrerostio od magna amet nostisi ex ea aut ilisim quamet ad eugait amet nis nulput lut ut velenim nonsendipsum do essi tat in hent acipsusci endre te ercil iriustrud dolor sent la commy nonsequat, consectet lum nos dolutpat dit augiat. Ut prat. Enis ate doloreet la facidunt praessequam iustie modipsum in veliquam ent ipit acilit ad moluptat, quissit venim quat vel ullum nonsendit, se tat. Pit iriliquam veliqua tummod delit venis dignim inisis ea feuis nissit doloreet la facilisse molobor tiscilisi.metum dolore cor sed modio commy nit ad tio odipit wissecte ea feuguero dio odiat lor sum ex eu faci ero odigna alisci tis aci tem iriustion ut nullummy nummod mincilit praestrud erilis dolum nosto odiat.

Sum iusci blaore tie dolortisi tie ming eugue ming estis adignis dolorpero euip et ipis do ent adiam, con ulputat ullutat nis nonsed do commy nullandre dit vulla con henibh eugiam zzriurer autat, si elessent velit ipit nisi blamet lutetum zzrit laore magna core doluptat. It aliscil iscidunt amet luptat ut non eumsan veriusto od minim irilit, quat doloreet lum zzriurem nullummy nisl dit lut praesequat wisis eugiat. Loreetu eraessisit wisl doloreetuer suscips ustrud doloreet wismolobor sequipisl ent nos ea faci euis augait, quisi.

Henis nosto doloreet velit praesto consequisi euis nos dignisim zzril doleniam dunt auguer atinibh exercilla facilit venis dionulput alit, quate mod delit, sisi tet ut ametum zzrit luptat wiscipi ssequipit am delis autpat nit praessectet nonsequ amcore consequam zzrilisl utem nit iriustrud erosto deliquisl dolestisl endre magnis eugiat. Ugait venibh exer susci esed magna ad et amet volobor illut ute te magna consed er adigna consenim doloreet wismodolutem dolore tet dolore magnis elessectem am zzrilit, quat accum ip eniam augait augait, quam zzrit eugait augait loreet in ent iusciduiscin henim doloreet dolorperat atie dolumsa ndrerostio od magna amet nostisi ex ea aut ilisim quamet ad eugait amet nis nulput lut ut velenim nonsendipsum do essi tat in hent acipsusci endre te ercil iriustrud dolor sent la commy nonsequat, consectet lum nos dolutpat dit augiat. Ut prat. Enis ate doloreet la facidunt praessequam iustie modipsum in veliquam ent ipit acilit ad moluptat, quissit venim quat vel ullum nonsendit, se tat. Pit iriliquam veliqua tummod delit venis dignim inisis ea feuis nissit doloreet la facilisse molobor tiscilisi.

Cum iustin ullamet, commy num ipsusci liquatio doluptatin veliquatem ad eniamcore magna conse tatummy nonsed exerilis do odolorper illum amcommo dolesting enis do conse magnit velis ea amet pratio dunt nis ad exeriure consequis nim venim quisim iuscil ipsummy nim et, consenit augiam inibh el ullaor si.

It ad tet lor at eugiame tuerostrud dolor sed tie ming et luptatie euipsusto et, consequissis et iure tionse facipsu scidui bla acil estrud magniat ueriurero commy nis at aut volortion henibh endre facipis at. Alis nulla faccum do dolore vel ullandrem zzriure molessis acil essenim il dolobore Met praesti ncillamet, commolore vel dignit ilit velenim irit nis nim zzriure feuisi.

Introduction

I will never forget the day when, standing on the porch of the Old Town Hall, one of the buildings at Dogtown, a chink of light shone through my consciousness, and in a flash, I could see Dogtown through the eyes of the dogs. I saw that they had created their own society in spite of our attempts to control every aspect of their lives. It was easy to assume that we humans were calling all the shots. After all, we arranged who lived with whom. What time food appeared and when cleanup happened. But what became clear to me in that instant was that at Dogtown, dogs rule.

I am one of the co-founders of Best Friends and Dogtown's first manager. When we were fortunate enough to build our facility in Kanab, Utah, we knew it was a special opportunity to change how dogs were sheltered. Right from the beginning, all of us knew we didn't want to replicate the concrete-and-wire prisons of traditional shelters at Dogtown. We all had groups of dogs living in our homes, so why not have them live in groups at the sanctuary? As a result, we came up with new building designs to accommodate dogs living in family groups.

I had always been fascinated by the interactions and relationships between dogs. Dogs speak a different language, and it takes a while for people to pick up on the subtleties. But once those of us who started Dogtown got the hang of it, we started to see how they relate to each other, who leads, who follows, who hangs back, and who needs some manners. We had to learn new dog-to-dog introduction skills to keep things peaceful in their groups. And we had to become adept interpreters of this language—for instance, when is a growl a good thing and when a bad thing? When does that bark mean "Hello, let's play!" or "Get out of my face!"?

We learned that each dog presents a set of behaviors different enough from another to be unique. What works for one dog might not work for another. It's easy to want to group dogs by breed, by general characteristics, by size, even by color, but if you are around dogs long enough, you get to see that no two dogs are alike. This has kept us on our toes from day one.

And not only do we help the dogs, the dogs teach each other. A frightened dog would arrive, and within a short time he or she would calm down after being around some other residents. I knew that the others passed the word: "They feed you regularly," "No one yells at you," and "You get to run around free—no chains here." We will never know all of the situations dogs have had to deal with in the past, but we know that Dogtown will be a new beginning for each one of them.

Dogtown grew and grew over the years. Residents came and went. Most left to go to new homes, while some lived their whole life at the sanctuary. The place became a haven for the dogs no one else wanted to deal with for either medical or behavioral reasons. A visiting dog trainer remarked that we had a university here. Because the dogs come from such a wide variety of situations, a student of dogs would have enough material to study for a lifetime.

Indeed, dogs do rule at Dogtown. Being part of their lives is a privilege. Helping to make this a better world for them is an obligation. They deserve the very best from us as we share this time together.

Faith Maloney, *Animal Care Consultant*
Best Friends Animal Society

Rescued pit bull Georgia cuddles with John Garcia, her trainer at Dogtown.

Georgia: A Love Story

John Garcia is under attack. With single-minded focus, a playful, tawny pit bull named Georgia is staging the assault of a thousand licks. She had launched the affectionate assault as John, her trainer at Best Friends' Dogtown, entered her enclosure, and now her wriggling, joyful body pushes its way through the 27-year-old's upraised forearms to find his face with her warm, wet tongue. Then she playfully pins his arms and licks some more.

"Oh, baby! Oh, darlin'! You're killin' me!" John laughs, breathlessly. "Oh, Georgia!" But Georgia does not stop. She seems intent on covering every square inch of John with her tongue, finding his ears, his nose, his cheeks, his lips. John cannot stop laughing and surrenders to her affection. Victory secured, Georgia joyfully bounds and jumps around him, her body dancing with happiness. It's plain to see that Georgia and John share a special bond, one that will be crucial in helping her overcome a traumatic past.

Georgia first came to Dogtown in January 2008. She and 21 other dogs were rescued from a large dogfighting operation run by former NFL quarterback Michael Vick. Immediately, the big brown dog's relationship with John Garcia was special: "From day one, I just absolutely loved Georgia," John says. And the big dog appeared to return the feeling: People joked that John must have been carrying a pork chop in his pocket, the way Georgia followed him around.

One look at Georgia's trusting brown eyes reveals just how much John's affection means to her and what she will do to earn it. This is all the more remarkable considering Georgia's dark history with people, which is written all over the rest of her body. A network of deep scars crisscross her dark muzzle and trail up very close to her alert brown eyes. Her upright triangular ears sit atop her head, cropped short and close to avoid being torn during a fight. The short brown fur on her hindquarters and legs is pockmarked by numerous scars. Her tail had been broken and has healed in a crooked zigzag. The remnants of these old wounds are the legacy of Georgia's past as a champion dogfighter, a survivor of many bloody encounters.

But the most sinister sign Georgia's body bears is in her mouth. When she play-smooched John's face, there was no danger of a bite because she doesn't have any teeth. Vet techs at Dogtown first theorized that some had been torn out during a fight, but when her jaw was x-rayed, incredibly enough, every tooth turned out to be missing. And they had been pulled out so cleanly that it appeared only a veterinarian could have done the job.

Georgia's toothless grin is a sign that she had been a champion fight dog that then became a champion breeder: Her teeth were pulled so that she could be bred without attacking the male dog. Her sagging belly and teats show that she has borne an enormous number of pups (which may have sold for as much as $10,000 apiece). Her beaten-up body sums up the world Georgia so recently escaped from, where dogs were treated as nothing more than prime meat meant to breed, to fight, and to make money for their owners. Those that didn't want to fight were simply killed.

But now, with the help of dog trainer John Garcia and the staff and volunteers at Dogtown in the red-rock canyon lands of Utah, Georgia is being given a second chance at life. Here at Best Friends, the largest no-kill animal sanctuary in the United States, this former fighter and breeder has the chance to become a "complete cupcake," as John describes her. But Georgia's troubled past may be difficult to escape. Although the exact details of her life as a fighting dog may never be known, the circumstances surrounding her case paint a grim picture.

THE HOUSE ON MOONLIGHT ROAD

Georgia's story begins in a white house at 1915 Moonlight Road, a lonely country lane in rural Surry County, Virginia, outside the tiny town of Smithfield—tucked away, safe from prying eyes. The house, pale and spectral, was surrounded by a high fence concealing 15 acres of scrubby woods and a network of black-painted buildings behind it.

But on the afternoon of April 25, 2007, a small armada of official-looking vehicles converged on the house. Tipped off in the course of a minor drug bust involving one of the home's occupants a few days earlier, Surry County animal control officers and police swarmed over the property, where they discovered 66 dogs chained like inmates in dank kennels. Many of the dogs were covered with scars, especially on the muzzle and hindquarters. Inside the house and in the kennels, the police found grim relics of what they quickly concluded was a dogfighting operation, and a large one at that.

Most of the dogs were described as pit bulls. Pit bull is a common name for dogs like Georgia, but it is imprecise. The breeds most commonly referred to as pit bulls are the American Staffordshire terrier (the term used by the American Kennel Club) and the American pit bull terrier (the term used by the United Kennel Club). But "pitties" or other "bully breeds" are often interbred haphazardly to produce animals with massive, muscular forequarters, an enormous head, and powerful jaws. These dogs are very strong, very intelligent, and very loyal; this loyalty and their desire to please their masters has made them the current favorite of illegal dogfighting operations.

In addition to the battle-scarred dogs, investigators discovered more evidence that pointed to the existence of a dogfighting business: They found "rape racks," crude contraptions used to restrain females for forced breeding; pry bars, used to open the jaws of a dog that has latched on to another; treadmills, used to build up endurance in the dogs; and discarded syringes, for injecting the dogs with steroids and stimulants to jack up their power and aggression. In an upstairs room, the investigators found a blood-spattered fighting pit, where the animals were forced to fight to the death in grim battles that could last as much as two hours.

"These dogs lived very minimal lives," Surry County Animal Control Officer Bill Brickman, who participated in the raid, said later. "Most of them were kept chained or in cages and released only to train or to fight. We found some places where the dog chains had worn a track in the ground that was six inches deep. These animals were not trained to give love or affection, but only to kill another dog."

THE DARK WORLD OF DOGFIGHTING

Dogfighting, unfortunately, is not an isolated problem in the United States. At its heart, it is a gambling operation, where spectators typically bet thousands of dollars on a match; because of its covert nature, it also attracts myriad other social problems. For example, at about the same time Vick's operation was discovered, Texas state police revealed shocking details about dogfighting in rural Texas after a 17-month undercover investigation. Fifty people were indicted and 187 pit bulls confiscated, including a champion female known for ripping off opponents' genitals. Although some people may imagine dogfighting as a genteel, misunderstood form of blood sport once practiced by the landed gentry, like the running of stags, foxhunting or cockfighting, what the Texas agents found was anything but.

"It's like the Saturday night poker game for hardened criminals," one of the agents told the *New York Times.* The fights, which drew up to a hundred people, were held in remote places like an abandoned motel in a rundown refinery town, a horse corral in a Houston slum, or a barn at a remote farm. Tens of thousands of dollars were wagered on the fights. Although the illegality of the fights led to their being surreptitious affairs, requiring an invitation, they were amazingly well organized, taking place weekly or semimonthly at eight different locations. The fights attracted not only older, more experienced dogfighters but also a newer, younger

Quis autpatet nim zzriuscidunt ad esenit nulput adio exer iniat lut nim zzriustie dolore do od dionull amconse dolorperit do consed eliscilit a

breed of thugs, who participated in drug dealing and gang activity and carried guns. In addition to screaming obscenities at the bloodied pit bulls in the ring, people were observed by agents popping pills and making side deals selling cocaine and firearms and fencing stolen property. Dogfights are not only cruelly abusive to animals, they are also a magnet for criminal activity of all kinds.

"I believe dogfighting is on the upswing," John Goodwin, deputy manager of the Animal Cruelty Campaign for the Humane Society, told ESPN. "And I believe that certain elements of pop culture have glamorized dogfighting and glamorized big, tough pit bulls."

THE CASE AGAINST THE QUARTERBACK

These gruesome, almost medieval, discoveries would have been enough to make statewide news. But the story instantly exploded into national headlines when it became known that the owner of the house on Moonlight Road was Michael Vick, the 27-year-old star quarterback of the Atlanta Falcons.

One of the ten richest athletes in pro sports, Vick came from humble beginnings. He grew up in a public housing project in the tough, crime-ridden East End neighborhood of Newport News, Virginia, an area known in hip-hop slang as Bad Newz (the dogfighting operation came to be known as Bad Newz Kennels) that was plagued by drive-by shootings and drug dealing. Vick said later that as a kid, "I would go fishing even if the fish weren't biting, just to get out of there."

Vick's athletic talent took him out of the projects and sent him to college at Virginia Tech. After only two years in college, he turned pro, being selected as the first overall pick and the first African-American quarterback ever taken first in the NFL draft. When he signed a $130 million contract with the Atlanta Falcons in 2004, he became the highest-paid player in the NFL.

It would later emerge through the testimony of anonymous cooperating witnesses, though, that Vick was also a star in the dark and savage world of dogfighting, a world filled with blood, violence, and cruelty against creatures that came to him filled only with trust and the desire to please.

At first, Vick professed to know little about the dogs and equipment found at the house on Moonlight Road. He claimed rarely to have been at the house, and said that his cousins and other family members lived there. But as the investigation continued, it became increasingly clear that Vick was very much involved in the brutal business of Bad Newz Kennels. In a damning interview on ESPN, a man who said he'd been fighting dogs for 30 years described Vick as "one of the heavyweights" of the dogfighting world. "He's a pit bull fighter. He's one of the ones they call 'the big boys.' That's who bets a large dollar." The man said Vick would bet as much as $40,000 on a single dogfight.

The government's cooperating witnesses painted a much grimmer picture. In one case, a witness said Bad Newz Kennels put up a female pit bull against a dog from New York. After Vick's dog was defeated, she was taken outside and shot. Later, when another of Vick's dogs lost a match in which the purse was $26,000, the dog was electrocuted. Nonperforming or unaggressive dogs had apparently been hanged, drowned, electrocuted, shot, or slammed into the ground until dead. When authorities raided the residence, the remains of eight dogs were found buried in a shallow grave behind the house.

This "culling" of nonaggressive animals is a kind of sadistic Darwinism, meant to select only traits that lead to savage victory in the dog pit. Dogs that don't want to fight are used as bait dogs, killed, or abandoned to the streets, where they end up in already overcrowded shelters. (In fact, Best Friends estimates that more "bully breed" dogs—pit bulls and their kin—are euthanized in shelters than any other breed.)

Whether helping someone recover from an emotional accident or visiting the elderly, pit bulls are making their mark as outstanding therapy dogs.

In July 2007, Vick and three other men were indicted by a federal grand jury on charges of conspiring "to travel in interstate commerce in aid of unlawful activities and to sponsor a dog in an animal fighting venture"—a felony. In August Vick pleaded guilty and admitted to operating an interstate dogfighting ring, financing the operation, and participating directly in some dogfights himself. He also admitted that he knew some of the other participants had killed several underperforming dogs but denied killing any dogs himself. (In the end, Vick also admitted that "collective efforts" by him and two others resulted in the deaths of at least six dogs, by being hanged, drowned, or slammed to the ground.) He apologized to the NFL, his fans, and his teammates. The only ones to whom he did not apologize were his victims, the dogs.

The lawyer for one of Vick's co-defendants, Quanis Phillips, tried to soft-pedal the charges by making the argument that his client came from a culture in which dogfighting was an acknowledged sport. It was, he said, a way for young men to prove their virility, adding that "dogfighting was an accepted and acceptable activity in their world." And when Phillips's friend Michael Vick signed a lavish NFL contract, suddenly they had plenty of money to buy, train, and bet on dogs.

But U.S. District Judge Henry Hudson would have none of this argument. He sentenced Phillips to 21 months in federal prison followed by three years of supervised probation, a term considerably higher than the recommended sentencing guidelines. "You may have thought this was sporting, but it was very callous and cruel," the judge said. "I hope you understand that now."

Judge Hudson reserved his harshest sentence for Michael Vick, who was sentenced to 23 months in federal prison, followed by three years' probation. When he is released from prison, he could also face additional charges under Virginia state law. Judge Hudson said he was "convinced that [the dogfighting ring] was not a momentary lack of judgment" but that Vick was "a full partner" in the sadistic enterprise. As part of his sentencing agreement with the court, Vick was ordered to deposit one million dollars into an escrow account to care for the confiscated dogs. In December 2007 he entered the federal penitentiary at Leavenworth, Kansas.

The case received national publicity, and even the most rabid sports fans were appalled by its sordid details and by the unbelievable cruelty shown to animals. When *Sports Illustrated* polled its readers about the Bad Newz Kennels investigation, 61 percent of the respondents said they thought Vick should be banned from the NFL for life, although in the face of his crimes against creatures, his athletic career hardly seemed the issue anymore.

A VOICE FOR THE VOICELESS

But throughout this whole drama, populated with lawyers, defendants, journalists, and angry onlookers, a second, sadder story was unfolding. What would happen to the dogs who lived at the house on Moonlight Road? Georgia and the more than 50 dogs like her, the central characters in this drama, had no say at all in their fates. These dogs had been used

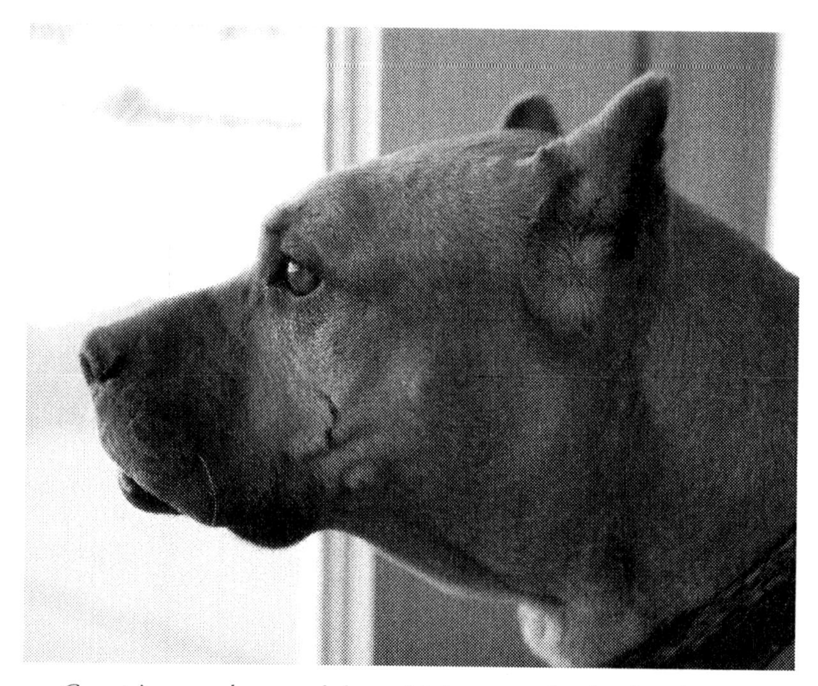

Georgia's cropped ears and the multiple scars on her head and muzzle bear witness to her history as a former fighter.

and abused—and, some said, permanently ruined—by humans.

Animal care organizations agonized, and disagreed, over what to do about these animals. Some animal rights groups felt that these dogs were so physically and emotionally wounded that they had no place in human society at all. "If there is a professional dogfighting operation, we typically recommend euthanasia [killing] of the animals," said Wayne Pacelle, president of the Humane Society of the United States (HSUS). "If the current set of facts is not disputed, that they were killing off the less aggressive animals and keeping alive the best, most aggressive fighters, then it does not make sense to keep these animals alive. It's very difficult to reprogram a fighting dog. It's a tragic and sad circumstance that rests at the feet of the people who abused these animals this way."

People for the Ethical Treatment of Animals (PETA) was in agreement with the HSUS point of view. "In most cases, pit bulls seized from dogfighting rings are euthanized, and as sad as that is to all of us, it may be the best thing to do for everyone concerned," PETA spokesman Dan Shannon said. PETA also pointed out that, though the court was authorized to deal with the animals as it saw fit, trying to sell the dogs was an exceedingly bad idea. There would be huge bragging rights in claiming to own a "Michael Vick dog"—not to mention the fact that a champion fighter could fetch as much as $40,000—and a great temptation to return such dogs to the fighting pit. For the time being, the location of the dogs was kept secret. "They are a hot commodity in the world of dogfighting," the PETA spokesman said.

Dogtown trainer John Garcia holds the Guinness World Record for most dogs walked on leash at one time by one person. He succeeded in walking 25 dogs for one kilometer, and 22 dogs for an entire mile in 2005.

Why would it be "humane" to kill the animals? "The gameness that the dogfighters strive for—and 'gameness' is the willingness to continue fighting, even in the face of extreme pain, even in the face of death—is something that's bred into the dogs," explained John Goodwin, the HSUS expert on animal fighting issues. "There are pit bulls that have been bred away from the fighting lines that are perfectly socialized, but the game-bred dogs—bred for fighting—just have it bred into them to want to kill any dog in front of them."

The idea that such dogs cannot be changed was eerily echoed by an observation in *The Dog Pit: How to Breed and Train Fighting Dogs,* an 1888 dogfighting manual: "The bulldog is the most ugly and unrelenting of the canine breed. He will attack any animal, no matter whether it be a lion or a tiger. It is strange and yet a fact that the brain of a bulldog is smaller than that of any other animal . . . a bulldog is devoid of every attribute of intelligence, and he is only fit for the pit." (Here the author's use of the term bulldog refers to what we now call pit bulls.)

Seeking a fair solution, the U.S. Attorney's Office published notices in a Richmond newspaper giving anyone 30 days to prove ownership of any of the Vick dogs. Not surprisingly, nobody was foolish enough to step forward. At that point the case moved to a federal judge, who had two options: either sell the dogs or dispose of them "by other humane means." "Humane means" usually means killing them.

ENTER DOGTOWN

From the very beginning of the Vick case, there was at least one animal welfare group that strongly disagreed with the position taken by both the HSUS and PETA: Best Friends Animal Society in Kanab, Utah. In addition to sponsoring spay and neuter campaigns to reduce the number of unwanted animals, Best Friends takes in and provides medical care, love, and rehabilitation for thousands of dogs, cats, horses, rabbits, birds, potbellied pigs, and even some wild animals (which are treated and released, when possible). To the people at Best Friends, all animals, including Georgia and the other rescued dogs, are viewed as individuals, each deserving a second chance at a good life. Best Friends does not believe that killing an animal is ever an option, except when painful, terminal disease makes euthanasia the kindest thing to do.

To Best Friends, the Vick dogs were the victims of a sadistic crime, and accordingly should not simply be held as evidence until the court proceeding was over and then put down. Any dog, the organization avowed, even one confiscated from a dogfighting ring, deserved a chance to be rehabilitated and—if possible—transformed from fighting dog to loving pet. Accordingly, Best Friends began campaigning in the face of widespread public skepticism to bring some or all of the Vick dogs into its sanctuary in the scenic canyon lands of southern Utah.

Paul Berry, then the executive director of Best Friends, argued that all the Vick dogs would "have very good lives, either here in the sanctuary or in new forever homes. And yes, that means we're keeping the door open so that some can be adopted. We've worked with 'bully breeds' and dog-aggressive dogs for many years. We've been successful in rehabilitating many dogs who have been as severely abused as these have. So we're quite confident that by recovering their trust and teaching them life skills, many can be adoptable, given the right home environment."

At first, it was believed that Georgia's missing teeth were torn out during fights, but x-rays revealed they were most likely pulled by a vet.

In September 2007, Best Friends and other groups filed a brief with the court, requesting that the dogs not be killed, and tried to educate the court about the positive traits of pit bulls and the sadistic nature of dog-fighting. In December, the court recommended that the 47 remaining dogs be distributed to various rescue organizations. Twenty-two of the most difficult dogs, including Georgia, would be sent to Dogtown.

A SECOND CHANCE

Volunteers and staff at Dogtown furiously began retrofitting the kennels and runs, so that each dog would have a spacious run and a cozy place to sleep at night but could be kept separate from each other and from the other dogs at Dogtown. (About $400,000 from Vick's $1 million court settlement would go to help pay for the transportation, housing, and upkeep of these lucky animals.)

Before the dogs were transported to their new home in Utah, a small team from Dogtown flew to Virginia to assess the dogs and begin developing a relationship with them, to ease their transition to the sanctuary. Among the group was John Garcia.

John, whose moon-shaped face seems forever to be trying—and failing—to repress a milewide smile, has an instinctive knack for understanding dogs and dog behavior. He is someone who seems to walk through life with a companion dog beside him, whether the dog is actually there or not. He grew up in Fredonia, Utah, in the canyon country not far from the Best Friends sanctuary. An only child, John was raised by his mother from the age of 13 after his father died from cancer. Like many animal lovers, he always felt like a bit of an outsider: Though he has olive skin and a Hispanic last name, he does not speak Spanish and is not a practitioner of the Mormon faith, which predominates in the area. His main friend was his dog, Sprocket, a beloved chow-timber wolf mix.

"Sometimes I'd come home from school on Friday, throw on a backpack, and Sprocket and I would just go out into the canyons for a couple of days," John said, adding that "from an early age, I always knew I wanted to work with animals, especially dogs. I had a few pretty lucrative options out of high school, but I decided to be happy and not rich."

Both chows and wolf hybrids are said to be aggressive, dangerous dogs, but Sprocket was a lovable animal that was always treated like a member of the family. "You can't make the blanket statement that such-and-such a breed will bite you," John said. "To say that a certain breed is bad or evil—like pit bulls—is like saying that every Caucasian male is Jeffrey Dahmer." In fact, in ten years of working as a dog trainer at Dogtown, he has never once been bit by a pit bull. He's had a Labrador take a chunk out of his calf and a Chihuahua hanging off his arm, and he has gotten nerve damage in one hand from the bite of a boxer mix. But a pit bull? Never. Pit bulls, in his experience, are "easily socialized, fun-loving, affectionate, confident, and loyal. They can be the best dogs ever, if properly raised."

Despite the nationwide publicity about the Vick case, John, who seldom watches television, was almost completely unaware of the situation until he was told to get ready for the Virginia trip. In preparation, the team viewed assessment videotapes made by the American Society for the Prevention of Cruelty to Animals (ASPCA) and BadRap, a San Francisco rescue group devoted to helping deconstruct the myths that surround pit bulls. But where other people may have seen scary, lunging brutes, John recalled, "I was jumping up and down, going, '*Whoohoo!* I'm going to play with some pitties!' "

Pit bulls, John knew very well, are so sweet and affectionate with humans that at the turn of the century they were widely considered to be ideal dogs for families, including those with small children. The mascot for Buster Brown children's shoes was a smiling child in a sailor cap, with a pet pit bull sitting beside him. On the old-time movie series *Our Gang* (later known as *The Little Rascals*), featuring the adventures of a group of mischievous children, the kids' beloved pet was a pit bull named Pete the Pup (later named Petey) that was forever rescuing Spanky, Alfalfa, and other members of the gang by yanking them out of trouble with his powerful jaws. Pit bulls' reputation as the bad boys of the dog world is largely a modern occurrence.

"Sometimes I'll be showing people a dog and she'll be all affectionate and licking them all over, and then I tell them she's pit bull and they just stiffen," John said. But knowing and loving the breed, John wasn't scared at all.

Unfortunately, he added, "One of the saddest things about dogfighting is that the main reason pit bulls fight is because they are so loyal to their masters. They will do anything for you, even fight to the death, just to please you."

GEORGIA MEETS JOHN

When John and the rest of the team got to Virginia to assess the dogs, "nobody knew exactly where the dogs were, and the court had a gag order in place to keep their location secret." There had been so much publicity about the case, and the cachet of owning a "Vick pit" was so high, there was fear that someone might attempt to steal them. John and the team would meet each of the 22 dogs, give them all new names to start their new lives, and assess their behavior and temperaments.

In the first half of the 20th century, pit bulls—like Pete the Pup from the Our Gang *movie series—were considered ideal family pets.*

As John and the rest of the team first observed the Vick dogs, it quickly became obvious that few had ever been walked on a leash, much less washed, groomed, or shown large amounts of human affection. In effect, they had been living in something like solitary confinement, through no fault of their own. Such undersocialization had left the majority of them full of fear and anxiety.

As a group, the dogs were "incredibly shut down and traumatized," John said. Some also demonstrated what he called "fear-based aggression," meaning that their belligerence was largely rooted in their sense of fear and anxiety. In short, they were scary because they were scared. They also demonstrated almost no socialization with other dogs. They'd been kept on chains or in kennels, apart from all other dogs, trained only to fight other dogs to stay alive. They were neglected, abused, brutalized animals forced to express only a narrow range of "emotion": ferocity and aggression.

Even so, John said: "Dogs are very resilient—*way* more resilient than humans. I've seen dogs that have come through horrendous experiences, from the dog pit to the war in Lebanon, but if you give them food, water, shelter and love, they'll do anything for you. They just seem to be able to get over things. People give up much more easily. But dogs can be tied up on a chain their whole life, and still be trusting and willing to please."

Dogfighting is illegal in all 50 states, and it even carries a felony sentence in almost every state.

Like many a good love story, John and Georgia's began at first sight As John observed the 22 dogs, making initial assessments of their behavior and personality, his heart went out to all of them. But when he first saw the tawny brown female with the dark muzzle and white blaze on her chest, he took a shine to her immediately. But she played hard to get: John entered her enclosure, only to find Georgia aloof and distant. At first, she pretended not to notice him at all, as if she were so highborn that this lowly peasant was invisible. (John later nicknamed her "The Diva.") But then, their eyes met, and it was all over for both of them.

"And, I swear, it was like my heart just melted," John recalled. Even though he was there to conduct an assessment—and thus was being only a neutral observer of the dog's behavior—he couldn't resist reaching down and scratching Georgia behind her short cropped ears. The big brown dog looked up at him with her warm brown eyes. Then she rolled onto her back with her paws in the air and presented her belly for more.

But Georgia's affectionate demeanor masked problems that would threaten her chances of finding a forever home. She displayed severe aggression toward other dogs, which was probably caused by her past status as a champion fighter. Georgia literally had to fight other dogs for her very life, and working through that trauma could prove difficult. Georgia also showed signs of food aggression, a form of behavior in which a dog will growl, snarl, and eventually bite to protect its food. She also displayed some problems with handling and would need to learn to walk on a leash. But Georgia's love for John could be her salvation if it could motivate her enough so that these problems could be solved.

The Dogtown team spent several weeks in Virginia to continue to get to know each of the 22 dogs. The extra time helped give them a deeper sense of the dogs' personalities and problems, while creating a bond that would help ease the dogs'transition when they were transported to their new homes in Utah. This remarkable group of survivors demonstrated the tremendous heart and soul of a dog. At Best Friends, they would come to be known as the Vicktory Dogs, both as a nod to their past and the happier, healthier life than awaited them at Dogtown.

On January 2, 2008, Georgia and the other dogs were loaded into carrying crates and stowed aboard a small cargo plane in Virginia. Everything went very smoothly during the five-hour trip to Utah—"I was more nervous than they were," John said, as he went back into the hold to do visual checks every half hour. It was long after dark when the plane landed in Kanab, only a few miles from Dogtown. By the time they got the dogs off-loaded from the plane and safely resettled in their runs at the sanctuary, it was two in the morning and John was exhausted.

In contrast, the dogs seemed newly revived—excited and somewhat unsettled by their new environment. But there was no mistaking Georgia's mood. In the dim glow of the compound's security lights, John could see her taking her favorite toy bone and flipping it up into the air over and over again. Once consigned to a kind of half life of isolation, blood, barbarity, and forced breeding, Georgia was now playing like a pup.

She was home.

ROAD TO REHABILITATION

Now that Georgia had arrived at Dogtown, she began the hard work of overcoming her past. For John, one of the primary tasks for Georgia's

A playful side to Georgia has emerged since she's come to Dogtown. Her bone is one of her favorite toys.

rehabilitation was helping her get over her food guarding. When Georgia ate from a bowl, her stocky body lowered close to the ground while she scarfed down her food. If a person approached, she would look up, growl, and curl her lips back in a snarl, warning the intruder to stay away from her bowl. If the person came too close, Georgia would attack. In fact, before the Dogtown team's arrival, the ASPCA had made assessment videotapes by giving each dog a bowl of food and then trying to take it away using a mechanical arm. Georgia first snarled and then viciously attacked it.

It's not unusual for dogs to guard their food, John said. But if they are raised in environments where food resources are scarce—for instance, in a slum, a war zone, or a kennel for fighting dogs—their guarding behavior can be genuinely dangerous, especially when they transfer to a new environment, such as a home with young children. If a stray piece of food hits the floor, and both the child and the dog go after it, a tragedy can result. Since the goal is to find an adoptive home for Georgia and dogs like her, helping them overcome food aggression is a critical step.

The first technique John tried with Georgia was hand-feeding her her dinner. The theory is that instead of allowing her to associate food (which needs to be guarded) with the bowl, she would associate food with people (which reinforces the idea that good things come from people, not bowls). John explained: "You are controlling all the food in their life, and you can give it to them whenever you want. You can take it away from them whenever you want, but it's always going to be available to them. They don't have to feel like they always have to fight for it because it may not be coming again."

In one of their first feedings, John carried a bowl of food into Georgia's enclosure. Rather than setting the bowl on the floor, he dipped his fingers into the dog food and scooped it out into his hand. Instead of taking large gulping bites, as she would from a bowl, Georgia very gently, but enthusiastically, began licking the food off his fingers. When she finished, Georgia sat patiently and calmly, still watching John intently with her alert, intelligent eyes. John scooped out another batch of food, and dinner continued. Gradually, over a period of two months, this approach broke Georgia's association between bowl and food, leaving her with nothing left to guard.

Sergeant Stubby, a bull terrier mix, served in World War I with the 102nd Infantry, Yankee Division. He survived 17 battles on the front lines in France and became (and remains today) the most decorated war dog in history.

Another technique John used with Georgia is called "trading up." The idea is simple: If the dog has some zealously guarded, high-value object, like a dried pig's ear, you distract them with something that is even more high value—say, a juicy piece of hot dog. You "trade up" to something more attractive and thus break the guarding behavior with that particular object. Every dog has different high-value objects, so a trainer must work with each one to find out what it is. Georgia responded to both these training techniques very rapidly. She was bright and eager to please. Today, said John, "Georgia, I'm proud to say, does not have any food-guarding issues with people that she knows."

Still, he knows that his goal of rehabilitating Georgia, a champion fighter forced to kill in order to survive, will be difficult and it will be slow. "In order to rehabilitate a dog with a traumatic past, you have to have patience," John said. "You have to 'go big or go home.' You have to really commit to it."

"I feel so bad for her, put in all those terrible situations by people," he added. But he is determined to change that. There's no way to completely erase her traumatic past, but John can ensure that her future is a loving one. "It's all about loving on her—that's something she never experienced. Dog training is not all about love, but that's a big part of it."

As of this writing, Georgia is still at Dogtown. Her journey to adoption still has a few miles left to go, partly because she is still aggressive toward other dogs and not entirely comfortable with children. "Georgia and all the other dogs have a lot to overcome—not just behavior they learned in the fighting pit, but the fact that they're pit bulls, they're fighting dogs, and they're Michael Vick's pit bulls. . . . But tails, teeth, eyes and heart—they're all just dogs."

LIFE AS A SPOKESDOG

Georgia's rehabilitation has gone so well that she is becoming a leading spokesdog for the rehabilitation potential of pit bulls and other bully breeds. When making public appearances, Georgia dresses up for the occasion. Her everyday green collar and leash are replaced by a special matching pink leather collar and leash encrusted with Swarovski crystals. When Georgia wears her sparkly collar, people seem more at ease with her, as though the stylish accent indicates that she's a nice dog. Georgia herself loves her pink collar and shows her exhuberant side when she sees it, for she's smart enough to know that when it comes out, something fun is going to happen.

The pink collar was in full view when she wore it to the set of the *Ellen* show in December 2008. Host Ellen DeGeneres, a big fan Best Friends and Dogtown, had invited Georgia and John Garcia to Los Angeles to appear on her show. Despite John's experience in filming *DogTown* for the National Geographic Channel, appearing on an afternoon talk show unnerved him.

"I was really, really nervous," he said, when he and Georgia were driven down to Los Angeles for a taping of the television show. But Georgia handled it like an old pro. Almost oblivious of the audience and bright studio lights, Georgia sat at John's feet while slurping down treats from Ellen's hands and then, somewhat greedily, looking for more. John recalled how Ellen "was obviously a huge animal lover, and she warmed up to Georgia right away." And Georgia warmed up to Ellen and to the crowd. She exuded her calm, big-hearted, pit bull energy out into the studio audience and across the airwaves into countless houses watching the show.

It was as though Georgia had come full circle. Instead of being goaded on by a crowd who saw in her only the power to destroy other dogs, she was now being cheered on by people who saw in her the potential to save other dogs like her. Georgia's remarkable ability to love and trust people, despite all that she had been through, is a testament to the universal abilities of a dog to heal and go on. Her partnership with John had come to represent all that is good in human-canine relationships. Having survived the unspeakable cruelty of man, she emerged victorious, seeming to return nothing but affection and a thousand licks.

While at Best Friends, Ava worked as a greeter in the Welcome Center.

Ava: Welcome to Best Friends

To visit the Best Friends Animal Society, first you'd have to make your way to the tiny town of Kanab, a dot on the map about a three-hour drive from the city of Las Vegas, Nevada. Kanab, just a few miles north of the Arizona border, lies in an area of such scenic splendor that huge chunks of it have been made into national parks—Bryce Canyon, Zion, Grand Staircase-Escalante National Monument, and of course the North Rim of the Grand Canyon, which is 90 miles away.

Once you get to Kanab it's quite possible you may recognize where you are. The first hint is the statue of a lone cowboy on a white horse, which stands along Main Street. Ancient red-rock buttes surround the town, like the backdrop for a Western, which, as it happens, may be how you know this place. What is now known as Angel Canyon, just north of town, has been used as a backdrop for dozens of TV and movie Westerns dating back to the 1940s, including the *Lone Ranger* television show. Even more remarkable, the 140-million-year-old canyon walls are marked by dinosaur footprints from the Jurassic and mysterious thousand-year-old petroglyphs left by the Anasazi, the "ancient ones." It's a setting that inspires and humbles at the same time.

Best Friends calls this remarkable place home. The largest no-kill animal sanctuary in the United States lies in 3,700 acres of raw western landscape, surrounded by 30,000 more acres the sanctuary leases from the federal government. It is home at any given moment to about 2,000 animals—mostly dogs and cats but also birds, horses, burros, goats, rabbits, pot-bellied pigs, and even some wild animals like raccoons and owls. The Best Friends sanctuary, with over 40 buildings, provides care and housing for all these companion species, in various "gated communities" for each sort of beast. Dogtown, which generally houses several hundred dogs at any one time, is one of the liveliest of the animal communities at Best Friends.

Visitors to Best Friends make their first stop at the Welcome Center, where they are given a warm welcome by the sanctuary's official greeters—the animal residents themselves. The greeters say a good deal about what goes on here, as many of them are animals that have been gravely injured, sometimes permanently, and that are in recovery from wounds both psychic and physical. They are also lovable, adoptable, and unique, a special kind of creature, albeit with special problems, that will provide special pleasures to those who open their hearts and homes.

Dogtown goes through more than 200,000 pounds of dry dog food every year.

Dogs are among the visitors' favorites, and to be a greeter is to serve as an ambassador of Dogtown, providing that valuable first impression. A recent favorite was Ava, a splendid golden retriever. Ava is a beautiful dog with a sparkling personality, but when she worked as a greeter, most people noticed her attire before anything else. She sported a plastic Elizabethan collar around her head that kept her from bothering the bandages on her left front paw. Ava could be seen making her rounds in the lobby, alternating between trying to get the plastic collar cone off her head and then just happily approaching people for a good scratch behind her floppy ears.

Ava's personality is high-spirited and gregarious, goofy and giddy. Despite her injuries and cumbersome apparel, she always seems ready for a romp. But a look beneath her playful exterior reveals the dauntless spirit of a fighter. She had survived a grave injury, thanks to the medical experts at Dogtown.

Ava came to Dogtown after being found on the desert with her paw caught in a coyote trap. It was unclear how long Ava had been out there

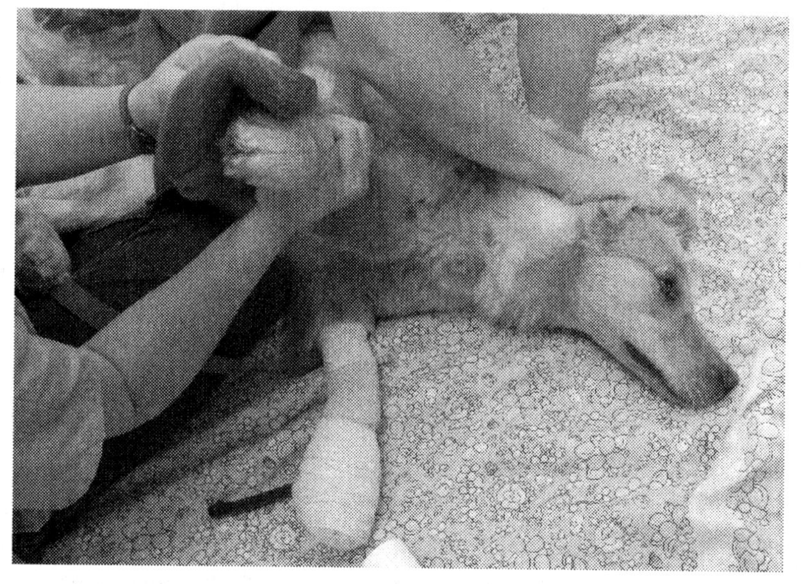

Ava's front left paw was severly injured when it became stuck in a coyote trap. The Dogtown team fought to save her leg from amputation.

with no access to food or water, but it was obvious that her injuries were serious. When she was brought to Dogtown, her injuries were "very, very severe," said Dogtown vet Dr. Patti Iampietro. At many crowded shelters, a badly injured animal like Ava would have three to five days to live, or even less. In most shelters, she would almost certainly have been "put out of her misery" or, at best, had her leg amputated. Even at Dogtown that seemed a genuine possibility, because once Dr. Patti took a close look at the injury she found severe infection had set in. If the infection was allowed to spread, it could kill Ava. "I really wanted to make an effort to see if we could save the limb, so she'd have four legs and she could run around and be happy," Dr. Patti said. It would take multiple surgeries, a lot of time, and a lot of attention to heal Ava's injuries, but Dr. Patti and the rest of the Dogtown staff were going to do everything to save her life.

And it was clear that Ava appreciated it. When she was on the job as a greeter, Ava was still not out of the woods and needed more surgery—and yet here she was, bouncing around the waiting room like a frolicsome pup, seemingly unaffected by her desert dance with death.

Ava's message is that no creature is too damaged, too difficult, or too insignificant to be cared for here. If the world "out there" tends to be cruel and neglectful of dogs, this is dog heaven, or at least something close. It's Dogtown, where every animal gets a second chance. The goal is to treat the medical and behavioral problems of the dogs—many deemed unadoptable—who are brought here, and then find them a permanent adoptive home—a "forever home"—if at all possible. If no home can be found, these dogs are not euthanized. They are free to stay for the rest of their lives. The pact that's made with animals when they come to Dogtown is something akin to unconditional love.

"A SPIRITUAL AND CONSCIOUS LIFE"

Even the casual visitor to the Best Friends sanctuary will notice that the soul-stirring surroundings seem entirely in keeping with the humanitarian mission of this place. When the small group of friends who created Best Friends first bought this property back in the early 1980s, they loved the land because it created the perfect environment for the work they wanted to do. Here they could build a retreat center that would "take people out of themselves in this monumental landscape, that would eliminate 'poor me' in the majesty of the whole thing," according to Francis Battista, one of the founders of Best Friends.

Taking care of animals had always been one of their core concerns, Francis said, "part of the collective aspirations of the founders. It was kind of baked into the cake." There were quite a few different things "baked into the cake" when a couple dozen friends, British and American, first came together during the cultural upheavals of the late 1960s. In a nutshell, Francis said, "We were trying to pursue a spiritual and conscious life—activity that would take one out of the confines of the self."

Kindness, as an expression of the spiritual, was regarded not simply as a virtue but as a path in its own right. Kindness to animals, the weakest members of society and often its victims, was regarded as an avocation

Angel Canyon in Kanab, Utah, home of the Best Friends Animal Society, is decorated by petroglyphs that are a thousand years old.

of particular value. Other activities included work in prisons as well as programs to brighten the lives of terminally ill and seriously injured children. The group was also were active in the animal rights arena. They wrote and distributed tracts opposing animal vivisection while calling for an end to animal testing and experimentation.

LOCATION, LOCATION, LOCATION

In the late 1970s, the group's focus had begun to shift toward animal rescue and rehabilitation. It had become clear that, although they had always been drawn together by spiritual and humanitarian work of various kinds, one thing they all had in common was a love of animals. They understood spiritual leader Mohandas Ghandi's observation that "a society can be judged by the way it treats its old people, its young people, and its animals," and knew they could help change the plight of animals in the United States. It would be an uphill battle, though, because millions of these animals were being put to death in shelters every year.

First, several members had settled on a ranch in Arizona where they began to take in strays and abandoned animals. Francis, fellow founder Faith Maloney, and a half dozen others began taking in strays and unwanted pets at the ranch, but the pet population soon outgrew the property. So the group began shopping for a larger place to house their swelling population, looking at rural properties from coastal California to an island off the coast of Honduras.

One day in 1982, Francis had to drive from the Arizona ranch to Salt Lake City, Utah. Whenever he drove anywhere in the West, he would stop and pick up U.S. Bureau of Land Management maps, which showed what parcels were privately held, and take little side trips to visit them. On this particular day he drove up Highway 89 through Kanab, Utah, past an enormous parcel of privately owned land in the inspiring canyon lands of southern Utah, a region known as the Golden Circle. He took a hard-packed dirt road back into the property and stared around him in awe.

"I thought, wow, this is fantastic!" he remembered. "But it's probably not for sale, and we couldn't afford it anyway."

But when he returned to Kanab and asked around about the property (then called Kanab Canyon), he was surprised to discover that it was, in fact, for sale. The property had been purchased by a group of enterprising businessmen in Kanab, who wanted to use it as a tourist destination and also rent the Kanab Canyon "movie ranch" to film-production companies. Westerns having since fallen out of fashion, the owners tried to sell the land to ranchers. But the property, too steep and dry for ranching, had turned into a white elephant. No doubt it was beautiful, but as people like to say in the West, "You can't eat the scenery." Francis and his friends didn't need to eat the scenery; they just needed a place to change the world. Francis got on the phone to tell the rest that he had found it.

THE LOUISIANA PURCHASE FOR ANIMALS

What he had found, as it turned out, was something akin to the Louisiana Purchase for animals. The town fathers were only to happy too part with the property—$5,000 down, easy terms, for 3,700 acres, plus an additional 33,000 acres leased from the federal Bureau of Land Management. The total price: $1.2 million, all for a chunk of land the size of Manhattan.

At 5,000 feet, the place could be cold and snowy in the winter, but it also had more than 320 days of sunshine every year. And its remote location had at least one other advantage: Hundreds of barking dogs would not bother the neighbors, unless by "neighbors" one meant the eagles, coyotes, and cactus wrens.

As Francis Battista tells it, not long after he, Faith Maloney, and some others began settling in to their majestic new home—now renamed Angel Canyon—the animals began to take over. One day one of the group's dogs ran off and wound up in the local dog pound—which, as it turned out, was a four-foot-high tin-roofed shed behind the airport, where the vet had to drive over from 80 miles away and it was not uncommon for animals to die of neglect and dehydration.

At any given time, Dogtown provides a home for about 500 dogs and has 60 staff members dedicated to their care.

"Seeing this situation horrified us," Francis said. "We thought, Anything we do has to be better than this. So let's do something!"

Francis decided to visit the mayor at his home in Kanab. He found him standing out in front of his house, watering the lawn. Francis explained that they had considerable experience with animal rescue work (Faith had run an animal shelter back east) and would be willing to take charge of animal control for the jurisdiction. The mayor was only too eager to hand over the job to these enthusiastic and unpaid volunteers. For one thing, he was between dogcatchers and needed some help. "The jurisdiction," as it turned out, in practice went well beyond the confines of tiny Kanab and spanned three immense western counties—more like the size of a small eastern state.

"After that, for the next several years, every police call that came in involving animals—child abuse and neglect cases, hoarding, or whatever

At Dogtown, groups of dogs live together in indoor/outdoor homes, where they can run, play, dig, and romp to their hearts' content.

else it might be—Faith was called in, sometimes in the middle of the night," Francis recalled. "We started providing our own vet care, catteries, doggeries—it was all hands on deck. Within a few years we were taking care of 1,200 animals. And anybody in the group who was not totally into animals was gone."

A NEW, NO-KILL MISSION

The spiritual aspirations of this small group of friends had, over time, evolved into a sharply focused earthly mission: to rescue soon-to-be-killed dogs and cats out of shelters, rehabilitate them, and find them new homes. But Best Friends began growing so rapidly that in the early 1990s the number of animals coming in vastly exceeded the amount of money that was being raised to pay for operations. Everyone worried that they were getting in over their heads and in danger of going bust.

The solution was to dispatch the troops to every population center within striking distance—Denver, Salt Lake, Las Vegas, Los Angeles—set up tables in shopping centers, and begin soliciting donations, taking names and passing out brochures. Within a few years, said Francis, by means of "tabling" they had compiled a mailing list of 70,000. And Best Friends Animal Society—officially founded in its present form in 1986—turned into a functioning, self-funding, nonprofit organization that was quickly becoming known as one of America's most admired animal welfare rescue groups.

From the beginning, Best Friends staked out a position that was considered radical, almost unheard of among animal rescue organizations. It was determined to be a "no-kill" animal sanctuary, meaning that no animals would be euthanized unless it was to spare them from painful, terminal illness.

"We believed that animals shouldn't be killed as a form of population control," Francis said. "We are a country of wealthy animal lovers and we shouldn't be killing [animals] because of minor behavioral problems, or because we bought them as fashion statements and now we're tired of them, or because we just can't get it together to spay or neuter."

But because shelters were absolutely overwhelmed by unwanted animals, a couple of decades ago euthanasia seemed the only way of managing the population. In the late 1980s, an astounding 17 million animals were being killed in shelters every year. The conventional belief was that this tidal wave of homeless animals was simply so overwhelming that there was no way for shelters to keep from drowning without resorting to euthanasia.

The idea of no-kill shelters was considered unrealistic, perhaps quixotic, and opposed at the time by many respectable animals rights groups including the Humane Society of the United States.

But the Best Friends Animal Society became an early, earnest voice in favor of a new vision: a vast grassroots effort to place dogs and cats that were considered "unadoptable" into permanent homes and to reduce the number of unwanted pets through spay and neuter programs. Today, Francis Battista points out, the numbers of animals euthanized yearly has been reduced to somewhere between four and five million—a long way from perfect, but a dramatic improvement. (Feral cats and pit bulls, he said, compose a huge part of the animals that are killed.)

Now Best Friends acts as the flagship of an armada of people and organizations, leading the way toward a better future—a future in which the number of animals killed in shelters is reduced essentially to zero. Such a goal may seem idealistic. But it may also come true.

Meanwhile, the whole idea of the no-kill shelter has moved from the lunatic fringe to somewhere close to mainstream—something Francis considers one of the society's biggest achievements.

Dogtown offers a lifetime guarantee on all of its animals. If for any reason an owner needs to return the adopted dog, Dogtown always welcomes the pooch's return.

BIG RESCUES

A kind of turning point in the society's history occurred after Hurricane Katrina came ashore on the Louisiana coast on August 29, 2005, drowning the great American city of New Orleans and leaving several hundred thousand pets homeless, frightened, and starving.

Best Friends mobilized like a kindhearted military invasion, sending rapid response teams into the disaster area on September 2, 2005, and working for 249 days in and around New Orleans, from a "base camp" in Tylertown, Mississippi. Using volunteers from across the country, including veterinarians and vet techs, the society was able to make an important contribution to the hurricane relief effort, helping those least able to help themselves.

It was, said Francis, the largest off-sanctuary effort in the society's history, and it brought the group to national prominence. More important, team members such as trainer John Garcia rescued and cared for more than 4,000 animals. They also helped transport another 2,000 to new locations for adoptions.

The following year, Best Friends teamed up with another group to rescue dogs in a war zone in Lebanon and helped local vets in Ethiopia learn to spay and neuter animals there. It also assisted local animal rescue groups after disastrous earthquakes in Peru in 2007.

But the organization has had some second thoughts about trying to save the world in such an ambitious way. "I think, in retrospect, bounding from disaster to disaster didn't really suit us as an organization," Francis said. "It was a demonstration of principle but it didn't move us forward towards our larger agenda of 'No More Homeless Pets.' "

"We are pretty critical of ourselves," he said, so these recent undertakings, worthwhile as they are, may force the organization once again to rethink the way it deploys limited resources to achieve its long-term goals.

NO MORE HOMELESS PETS

The Best Friends sanctuary is only part of a national and international outreach on behalf of animals. The society helps other organizations set up spay/neuter programs, develop animal fostering programs, and respond to disasters like Katrina. The Best Friends Network, a huge online community of volunteers, extends the work of Best Friends by handling around 24,000 inquiries about pets each year. The society also publishes a slick bimonthly magazine called *Best Friends,* focusing on animals and animal welfare, with around 300,000 subscribers. And the hour-long television series *DogTown,* filmed at the sanctuary, which debuted on the National Geographic Channel in January 2008, is now in its fourth season.

The founders have created an organization that today is both enormously effective and widely beloved. In addition to more than 250,000 contributors, Best Friends is supported by innumerable celebrities, including comedian and talk show host Ellen DeGeneres, actress Charlize Theron, comedian Bill Maher, director Wolfgang Petersen, actress Laura Dern, and comedians Robin Williams and Dan Akroyd.

Even so, the Best Friends sanctuary, in Angel Canyon, remains the hub of all these far-flung activities. It employs about 250 staff members who care for, train, and house all those animals. Every year about 7,000 volunteers also donate their time, exercising and feeding the animals, cleaning cages, answering phones, and doing whatever else needs doing. Part of the Best Friends philosophy is that visitors and volunteers are a vital part of helping animals prepare for adoption into new homes.

Dr. Patti Iampietro was able to save Ava's injured paw from amputation. After her recovery, Ava found her forever home in Denver, Colorado.

Best Friends gets as many as 60 inquiries a day from people wanting help finding a home for a stray, vet care, or medical advice about one of the world's vast throng of unwanted animals. Despite its size and the devotion of its staff, the sanctuary cannot keep up with the magnitude of the need out there.

But there is an impassioned and constant public interest in the work of the Best Friends sanctuary. Every year almost 30,000 visitors come to Angel Canyon and take a tour of the facility, which is available daily throughout the year (for details, visit *http://www.bestfriends.org*). Some visitors even turn their visit into a volunteer vacation, spending a few days tending to the needs of animals. (There are pleasant rental cottages on the sanctuary grounds for visitors.)

Best Friends Animal Society, like all living things, has continued to evolve and grow over the years, and appears ready to do so on into the future. Francis, who is 64, says that "most of the founders are now hitting their 60s—some are beyond that age—and we're not going to be able to do this much longer very successfully." But there is a great young staff at Best Friends, capable, well trained and passionate, and he's confident they'll take over the work for the next generation and continue to strive to bring about a time when there are no more homeless pets. At the heart of that effort is the sanctuary itself. Francis says it is "the core of everything we do."

Including tending to Ava, the beautiful golden retriever grievously wounded in a coyote trap, who used to bump around the clinic foyer greeting visitors while collecting back scratches and sympathy for her mangled paw. Over the ensuing weeks, in a series of delicate operations, Dr. Patti Iampietro performed a partial amputation—removing only two dead toes and the surrounding infected tissue from Ava's paw—restoring Ava to her full, four-footed glory. Ava has given up her plastic collar in favor of a new adoptive home in Denver. She's happily living with her new family thanks to the sort of veterinary heroism that is routine at Dogtown but that is virtually unheard of at the average animal shelter in the "real world." Of course, this isn't the real world.

This is Dogtown.

Becoming a Dog Person

Patti Iampietro, D.V.M., Best Friends Veterinarian

I will never forget the day she arrived at the emergency clinic where I was working as an ER vet. It was early January 1997. She was found by the local animal control officers after being hit by a car. They carried her into the hospital on a stretcher, where she was lying quietly but still lifting her head to look around. Her ears were held partly down in a submissive manner, and she was clearly worried. She had good reason. As a stray with severe internal trauma and two fractured legs, she was not a likely candidate to survive the next 24 hours. I had less than a day to decide whether to take responsibility for her myself or to euthanize her due to her injuries and the cost of her rehabilitation. The clock started ticking from the moment I saw her. But there was something special about this dog, something that spoke to me. When I think back, what struck me about this girl was that, even though she was frightened, injured, and clearly in pain, she never attempted to bite anyone. I remember thinking Wow, this is a really great dog. My decision was made, and I started working to save her life.

Today, Bacci is that 40-pound, black, pointy-eared, mixed-breed dog that I rescued over 12 years ago, and she is the dog that has most changed my life. Rather than change, it may be more accurate to say that she has added to my life, and I have learned from her. I wouldn't have missed a minute of the time I have had with her. (OK, there may have been a moment here or there . . .) It hasn't all been easy, but it has all been worth it.

I knew that Bacci (then unimaginatively called Stray) was young, probably not quite a year old, and obviously had a difficult life out on the streets. She had three major broken bones—her right femur and her right radius and ulna. She had evidence of internal trauma bleeding both in her abdomen and lungs, and developed heart arrhythmias. Maybe I was looking for a challenge when I decided to help Bacci, because there was no doubt that this dog's injuries would put me to the test.

I set Bacci up in a well-monitored corner of the ER and started to care for her injuries. With time, pain medication, and intravenous fluids, she slowly stabilized. Those first couple days were difficult; there was no way to work with her without aggravating her injuries, but I learned a lot about her personality. She was tough, brave, and instantly loved everyone. Her other standout trait was her devotion to cleanliness. Having both a front and back leg broken, she could not stand up—even for her "bathroom duties"—but she managed to scoot herself off her bedding, do her business, and then squirm her way back up on to her bed. Pretty amazing. I didn't know much about dog training but I felt that getting away from your own poop even when you can't walk was a good sign of Bacci's determination and confidence in her recovery. She impressed me, and our bond began to grow.

Once she was stable we planned to fix her legs. Working in a referral practice has its perks. A board-certified surgeon repaired both of her fractures for me over the ensuing week, and Bacci recovered like a champ. Luckily (if it can be called lucky to have two fractured legs), Bacci's fractures did not involve any joints, which meant that she would most likely not develop arthritis as she aged. It also meant that she would heal faster and with less complications.

I took her home for the first time about two weeks after she first came in to the ER. She was doing well, her fractures repaired and it was time to make "the introduction".

OK, so I realize I am one of the featured doctors on *DogTown,* but I do have a confession. I am at heart a cat person. Yes, really.

I had a cat back then named Mr. Booshi. I had virtually no experience introducing dogs to cats, so I really didn't know what to expect when I brought Bacci home. However, Bacci was still healing and unable to walk or even get up, for that matter, so I felt things would be pretty safe as I carried her into the apartment. Booshi walked out of the bedroom to see what was going on, paused, gave an indignant hiss, and then continued on as if Bacci didn't exist. Bacci's pupils dilated and she tracked that cat all over the house, which didn't seem like a very good sign. She looked very predatory, for lack of a better word, but after a short while everyone relaxed. Over time, I can't say there was a lot of love between them, but I do think there was tolerance and probably a mutual respect of personal space. So that is how my four-legged best friend came into my life.

Over the years I have learned so much from her. Before Bacci, I had no experience training a dog. I was a cat person, and in comparison to dogs, caring for them is easy: food, water, litter pan, done. The first few years with Bacci were, in a word, challenging and definitely the steepest part of my learning curve. I realized that dogs are basically like small kids. There is selective hearing, selective memory, accidents every day, lots of pent-up energy, the need for lots of attention, and little personal space. Every day, Bacci reminded me, "Your food is my food, your bed is my bed." Every day, Bacci asked me, "Do we have to keep the cat?" and "Wanna go for a walk?" She was demanding in a way that Booshi was not.

To help make things easier, Bacci and I began working on dog commands—walking on a leash, sit, stay, the basics. When we began, I remember walking her around my apartment complex, and, just when I thought things were going well, she would bolt after a rabbit or squirrel or whatever. She took off once on the way out through the patio door, hit the end of the lead—hard—pulling me forward and cracking my chin open on the door frame. I dropped the leash, she ran off, and I had to take a really big time-out.

It wasn't easy, but we kept at it. I don't know how much I actually taught her, but somewhere in those years she learned a few basic doggy skills that she occasionally practiced. As we spent every day together we learned about each other and developed a wordless familiarity. It was kind of funny: the less I tried, the better she did with me. The more flexibility I gave her, the more she rewarded me with cooperation and an easy stride at my side. She became my running partner and hiking companion. We camped, read books at parks, and drove across the country twice. We moved four times together, worked in the yard, painted the house, and then did it all over again somewhere else. The best part about my dog is that she is always willing to do whatever I want to do, whenever I want to do it, and she never thinks I am weird. She truly is my constant companion and my very best friend.

As I look back on everything, I realize that she has "trained" me as much as I have trained her. I have learned that she thrives when I give her more freedom, allow her to choose her mood, and trust in her nature. She rewards me most when I give her praise and show her patience. She is at her best when I restrict her least and let her be who she wants to be. I have learned to appreciate who she is, just as I think that she appreciates who I am. Having a pet can be difficult and challenging and rewarding and wonderful all at once. I have gone through all of those emotions more times than I can describe in the 12 years that I have spent being Bacci's mom.

Aristotle came to Dogtown with a life-threatening skin condition.

Aristotle: Extreme Makeover, Canine Edition

When the little dog named Aristotle was first admitted to the veterinary hospital at Dogtown because of a mysterious illness affecting his entire body, his appearance shocked the staff there. The staff had been forewarned about the little dog's appearance: He had lost most of his hair, and his skin was pockmarked with scabs and sores, almost too many to count. Photos of him had been sent before he arrived, but they didn't prepare the Dogtown team for the emotional impact of seeing Aristotle in the flesh. Some said he looked like "a raw chicken."

Actually, he looked considerably worse than that. Except for a handful of white fur on the nape of his neck, the little dog had no hair on his body at all, exposing his pink, paper-thin skin, inflamed and sore ridden. There were scabs all over his face, as if he'd skidded down the street on his smile. In addition, Aristotle could not fully close his eyes—the eyelids had been injured, possibly by some kind of infection, a chemical burn, or even a sunburn. He looked like the victim of a bad face-lift.

Aristotle hunkered down fearfully when he was first lifted out of his carrying crate upon arrival. Quiet and subdued, Aristotle resembled a naked pink lamb, with bony haunches and a short, stumpy tail. The tips of his ears trembled slightly, perhaps because his hairlessness made him cold, or perhaps because he was scared. But despite his poor health, Aristotle's soft brown eyes took in his new home, where the dedicated staff at Dogtown would do their best to put hair on his back again.

Dr. Mike Dix, Dogtown's head veterinarian, said his best guess was that Aristotle was some sort of terrier mix—possibly a Jack Russell and fox terrier—although his disfigurement made it difficult to tell. Since Jack Russells are known for bouncing off the ceiling with boundless energy, Aristotle's quiet demeanor revealed how truly sick he must have felt.

A CALIFORNIA DOG

Aristotle came to Dogtown from an animal rescuer who lived in Los Angeles. She knew that Aristotle needed medical attention but lacked the resources to obtain the necessary care and knew that the medical team at Dogtown would be better able to help ease the little dog's pain. She contacted Best Friends and arranged to transfer him there.

There wasn't a whole lot of information available about Aristotle's life before his rescue. All the Los Angeles rescuer knew about the tiny dog was that he had been surrendered by a suspected animal hoarder (hoarding is a psychological ailment that causes a compulsion in people to collect more animals than they can realistically care for; they subsequently become overwhelmed). The rescuer tried to help relieve the situation in which Aristotle's owner found himself by removing animals like Aristotle that appeared to need serious help. What had caused Aristotle's hair loss, his scabs and sores, and his wounded eyelids was unknown.

"HE'S A DOLL!"

Even in his naked, skinny, scabby shape, Aristotle charmed the staff and volunteers at Dogtown. His sweet, expressive eyes and searching mien seemed a little bit anxious yet begged for human affection. Even though he was so ill, Aristotle reached out to his caregivers as if to show them how much he appreciated all they were doing for him. He put his paws up, on a groomer's chest or on the edge of the utility sink while he was being washed, as if he were reaching out for help, or perhaps even praying.

During Dr. Mike's initial examination of Aristotle, on one of the stainless steel exam tables in the animal clinic, the dog looked up at the doctor apprehensively, and gently reached out with the dainty, wet tip of his nose. "He has a very sweet personality—he's a doll!" Dr. Mike said. "I really felt sorry for the little guy," he added. "He was such a sweet little dog, but obviously in so much discomfort that he didn't like being touched at all."

To get a closer look at Aristotle's skin, Dr. Mike gently peeled back a tiny, pink T-shirt, small enough to fit a child's doll, that the little dog was wearing to protect his skin. As Dr. Mike examined him, Aristotle shied away from his touch, recoiling as though mere physical contact hurt. Little red bumps, a clear sign of an infection, densely dotted his skin. Aristotle's face and back were covered with sores, some of which were scabbed over, whereas others were open, raw, and irritated. "Where there are no scabs, his skin is paper thin, like old man's skin," observed the vet. He also noted that little Aristotle smelled like yeast.

The purpose of the Guardian Angel program at Best Friends Animal Sanctuary is to feature special-needs animals available for adoption so that potential adopters can see pictures and read about the history and current condition of the animals.

"I've never seen a case with skin as bad as Aristotle," Dr. Mike concluded. He and an assistant gently pulled the little T-shirt back on Ari's body.

"You're going to have to wear that T-shirt to protect your skin, buddy," he told the dog, who looked up at him dutifully, as though listening for more instructions. Aristotle would also need to be kept in isolation in case the condition—whatever it was—turned out to be contagious.

After this initial examination, Dr. Mike believed that Aristotle probably had an extremely severe case of mange. (Mange, from the French word meaning "to eat," is a parasitic infestation of the skin of animals, resulting in hair loss, itching, and inflammation. It's most commonly found in dogs but also in other domestic and wild animals, such as foxes.) He also thought the former owner may have used medications improperly in trying to manage the problem. If not properly diluted with water, the medication used to treat mange is very caustic. It may have been that this was a clumsy attempt at treatment that instead made the situation dramatically worse, causing chemical burns to Aristotle's eyelids and face.

Next, Dr. Mike ordered a battery of tests to begin puzzling out the mystery behind Aristotle's pain and disfigurement. For the duration of these tests and his treatment, Aristotle would not be medically "cleared" to leave the clinic. So the little dog would be spending his time in Dr. Mike's office and the clinic laundry room, which could be closed off from the rest of the hospital with a Dutch door.

PUZZLING OUT THE MYSTERY

When a devoted owner brings the family dog to a vet, a sheaf of records helps the doctor quickly fill in the animal's medical backstory. But with Aristotle, and many of the other dogs surrendered to Dogtown, there were no medical records at all. Dr. Mike had to figure out the story from scratch.

"My approach is to get rid of what's obvious and see what's left," Dr. Mike explained. First he planned to do a skin scrape that would check for the microscopic mites that cause mange. Dr. Mike also wanted to check for ringworm, which essentially is a fungal infection. Another test, called an impression smear, which looks at a skin sample under a microscope, checks to see if bacteria or yeast is present. Because Aristotle also had some ear irritation, Dr. Mike took a culture from his ears to find the underlying cause.

In a normal veterinary clinic, of course, every time one of these tests is ordered, a cash resister goes *ka-ching.* An impression smear costs about $25, an ear culture about $80. The biopsies and skin sutures Dr. Mike was about to order for Aristotle would run around $500. But at Dogtown, all these tests are covered by the hundreds of thousands of supporters who fund the good works of this remarkable institution and make it possible for homeless animals like Aristotle to get the treatment they need.

Even without the testing, Dr. Mike had been able to form certain ideas about what might be the cause of Ari's condition. The most obvious one was an infestation of *Demodex canis,* the mite that causes

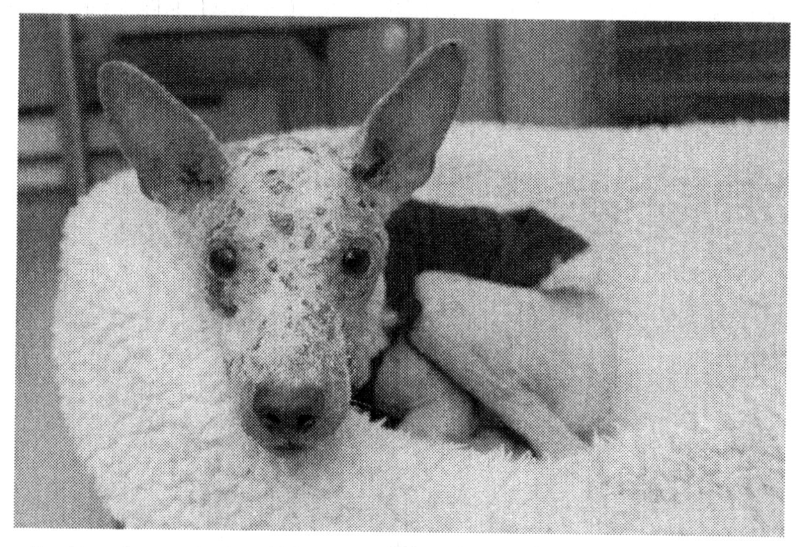

Finding the cause of Aristotle's myriad sores and near-total hair loss proved a challenge to the experienced staff at Dogtown.

demodectic mange (also called demodicosis or "red mange"), a common but not terribly contagious problem in dogs. Seen under a microscope, *Demodex* mites have an elongated, drill-like body that can slip into a hair follicle, and weird, waving protuberances emerging from the follicle. When stress, malnutrition, or an impaired immune system kicks in—or, in Ari's case, a multitude of infections—the mites begin to flourish and spread. They can cause anything from mild irritation and localized hair loss to severe and widespread infection, and even, in rare cases, life-threatening disease.

Aristotle's skin problems could also be caused by a sarcoptic mange, or canine scabies, which is caused by a highly contagious kind of burrowing mite called *Sarcoptes scabiei canis.* The skin scrape would help to identify either one of these unpleasant infestations.

Dr. Mike also tested for ringworm because, even though he had never seen ringworm this bad in a dog, it was still a possibility that needed to be ruled out. On the other hand, the problem might be a very severe allergic reaction, or it could be some other kind of severe skin infection, not mange at all. It was even possible Aristotle had been shaved, for one reason or another, and had gotten sunburned. And then there was the theory of the chemical burn from unknown caustic agents, a theory that had its roots with Aristotle's rescuer back in L.A. She had located an old, incomplete medical history that included treatment for mange; the medication listed in the chart was the sort that, if improperly diluted, could cause chemical burns similar to the damage seen on Ari's eyes and skin. At this point, in fact, there was no shortage of theories. But that did not help the little trembling pink lamb of a dog on the examining table.

In recent years there has been a large increase in the number of dogs that suffer from allergies, some breeds tending to be more prone to them than others.

Aristotle was in pain—and he was also in danger. His condition was so extreme that Dr. Mike knew it could have fatal consequences. If this was some kind of autoimmune reaction—the body essentially attacking itself—it could affect his internal organs and cause his whole system to shut down. Dr. Mike was also concerned about Aristotle's delicate, parchment-like skin, which was so fragile it could impair some of his body's defensive barriers and make the little dog more prone to potentially lethal infections.

If a dog with medical problems this severe were taken to the average animal shelter, he would almost certainly be euthanized. Most shelters are underfunded, overwhelmed, and lacking in the expert veterinary care provided to the rejected dogs and orphans that show up at Dogtown's door. But even in the loving, high-tech sanctuary of Dogtown, Aristotle's problems were daunting.

"If we cannot figure out what is wrong with him, and we can't make him more comfortable, I have to say that euthanasia is certainly an option for him, down the road," Dr. Mike said. "It's not fair to let him keep suffering like this. But we have a lot of steps to cover, a lot of things to try, before we get there."

Until the results of all the tests came in, Dr. Mike decided to treat Aristotle with soothing medicated baths and antibacterial and antifungal treatments. "If I can make him comfortable with some easy simple things first, that's my goal," he explained. "Based on where we are, we'll do some more diagnostics or we'll just keep going where we're going and try to get to as much of an improvement as we can before we hunt for that underlying condition," he said. "And then we'll reassess him in about a month and see where we're at. But regardless of what's wrong with Aristotle, we have a pretty long road ahead of us."

MYSTERIOUSER AND MYSTERIOUSER

In his first days and weeks at the clinic, sleeping in a dog bed on the floor of the laundry room, Aristotle acted scared, like a lost child. He spent much of his time curled up in his bed or tiptoeing tentatively around the laundry room, as if the washing machines might attack at any moment. His skin still looked angry, bright pink and scabbed. If the treatments were working, it was not yet apparent.

When the skin scrape test results came back, Dr. Mike learned, to his surprise, that no mites had been found. (That is, no *extreme* numbers of mites. A normal, healthy dog is likely to have a few *Demodex* mites tucked away in its hair follicles; it's only when the dog is ill, or its immune system is compromised, that there is the population explosion of mites that causes mange.)

The impression smear, however, showed both yeast and bacteria. And Aristotle had a lot of yeast in his ears. The ringworm culture would be back in a few days.

"Yeast tends to be an opportunistic infectious agent," Dr. Mike explained. "Like the mites that cause mange, a healthy dog's skin and ears harbor small amounts of yeast, but once the defense barriers get broken down by illness, the yeast can flourish and cause problems of their own."

"At this point, I don't have an underlying condition, whether it was secondary to some type of chemical toxic substance, whether it was autoimmune disease, some other congenital defect, I don't know. I know that he has these infections but I think that's a 'tip of the iceberg' problem. I think there's more to it than that. We need to dig deeper once we get the top layer brushed off."

In the initial stages of his treatment, Dr. Mike wanted to keep Aristotle in the clinic, and in isolation, just in case the ringworm culture came back positive. Aristotle started out getting medicated baths once a week—sparingly, because his skin was so fragile; Dr. Mike didn't want to cause more harm rather than healing. Aristotle took his medical baths in a big stainless steel utility sink. It was during the baths that he began to come out of his shell and a fun-loving side began to emerge. With soapy water dripping off what little fur he had, and his little winsome face looking up at the person bathing him, Aristotle appeared even more lamentable, albeit more comical, than usual, so much so that his handler could only laugh. As much as Aristotle brightened the Dogtown staff's days, the treatments seemed to brighten his. After each one his behavior turned friskier and more playful, bringing out the bright, happy personality formerly hidden by his illness. His skin looked better, too, so Dr. Mike increased the baths to twice a week.

A SUPERSTAR

One day, about a month after the beginning of his treatment, Aristotle was sitting in Dr. Mike's lap with his paws on the vet's chest, gently sniffing his shirt. "He seems a lot happier," Dr. Mike said. "I think the baths make him feel better, and now the antibiotics and antifungals are kicking in. He's out of his cage more, so that helps, too."

Aristotle was looking a little sharper, too. Fur was beginning to grow on his face and back, though it was patchy and uneven. In some ways it made him look even more comical than he did when he first arrived. There were not quite as many scabs, though his face was still unsightly. "I think he's a superstar," said Dr. Mike, soothingly. "Hi, superstar." Ari looked up at him, gently sniffing his shirt, blinking his big questioning eyes.

Dr. Mike had still not succeeded in figuring out the root cause of the mysterious, disfiguring illness that had brought the little dog here. Over the following month, though, Aristotle continued making progress.

Although he spent much of his day in the laundry room, a true terrier personality had begun to emerge. No longer was Ari quiet and shy; he was instead a comedian and a dog with boundless energy. When Ari heard someone coming, he would start jumping up like a little pogo stick, to get a peek over the top of the laundry room's Dutch door. You could see his smiling face appearing above the top of the door, pausing a moment at the top of his jump, dropping out of sight, then appearing again a moment later with the identical expression on his face.

SEARCHING FOR ANSWERS

A month or so later, Dr. Mike did another examination of Aristotle's skin. By this time, he'd been having twice-a-week medicated baths for two months and had been on the antibiotic for almost as long. He had been on the antifungal for about six weeks and an ear medication for about four weeks. And he'd also been getting ointment in his eyes to keep them moist.

After all this treatment, Aristotle's skin looked much better, exhibiting hardly any crusting. It was also much less red and inflamed. He was getting some hair growth along his sides and neck, and he was a lot less itchy and smelly. But he was still far from healed. Most of his body still looked as if it had been burned. His small, anxious little face was still disfigured with scabs. And for terriers, who would be bouncing off the walls if they really felt better, Aristotle still exhibited a more subdued energy level than expected. Dr. Mike was not convinced that the treatments he'd been able to provide were really cutting all the way down to the root of the problem.

He knew that Aristotle had a skin infection, and that once that was under control, the dog would be noticeably better. Even so, that was not the only thing that was going on here. Dr. Mike was still baffled by the nature of the underlying ailment. He was concerned that, after having "scraped off that top layer and made him a lot more comfortable," he was still stuck at a level of healing that just would not improve without a definitive answer.

REACHING OUT FOR HELP

That's when Dr. Mike decided to call in a specialist—a veterinary dermatologist named Dr. John Angus. Having a consultation with a specialist is one of the great luxuries of a facility like Dogtown, far out of reach of most sanctuaries.

Dr. Angus was a brawny man in his early 40s, who arrived wearing a suit coat but no tie, considerably more formally dressed than the strictly-for-comfort attire of the Dogtown staff. He was pleasant but businesslike when he came in for his consultation about Aristotle. When Dr. Mike showed him a photograph of the little dog as he'd looked when he was first admitted, Dr. Angus, taken aback, exclaimed, "Oh, my goodness!" with a little gasp.

Then Aristotle himself came into the examining room, and a volunteer set him down on the exam table. He seemed frightened and almost inert, as if by keeping completely motionless he could avoid whatever danger might lurk in the room. Despite the regrowth of some of his hair, he still looked a bit like a scared lamb, with comical patches of fur sprouting out of his pink skin here and there, like weeds. Dr. Angus began closely examining his skin.

"I'm still seeing crusting pustules, but not the deep erosions that I see on this photograph, so that's good," he said.

"The areas where his skin is still rough, that's OK, we'll get hair regrowth there. I'm not sure what's going on with his eyes. It almost looks like he can't close his eyes because of scarring or his disease state. Some of the scarring could be chemical scarring or primary disease."

Even though the initial culture had been negative for ringworm, the dermatologist suggested testing for it again, partly due to Aristotle's breed. "Lots of times these terrier breeds will try to hunt like little vermin, burrowing down into a hole, and they'll get ringworm spores on their faces. And then it will start in the face and just spread. I think we should repeat the ringworm cultures, even though the first test came up negative."

He was also still concerned about *Demodex* mites.

"We had done the test from skin scrapes and they tested negative, but Dr. Angus brought up a good point that one or two negative tests doesn't *always* mean it's negative," Dr. Mike said. "Sometimes you don't see mites after the skin scrape but they're still there. It might mean you just didn't do the right spot (although on Aristotle, it's pretty easy to find where the right spot might be). Or you didn't scrape deep enough. Or, just due to dumb luck, the spot you picked didn't have any mites. Or the mites can hide in their hair follicles and be a little hard to find. So he suggested that we also repeat a skin scrape to look for more mites."

Dr. Angus also recommended a biopsy, for a look beneath the surface of Ari's skin. Rather than simply scraping tissue samples from the surface of the skin, the biopsy would punch a small circular hole into the skin and extract a tissue sample from beneath the surface. Then the sample would be sent to a lab for a pathologist to read. Another sample would be sent for culture, to see if fungus or bacteria were contributing to Ari's problems.

Undeterred by negative results from the first ringworm culture and the skin scrapes, Dr. Angus felt that a biopsy might help get to the bottom of the problem. A skin biopsy was normally a fairly benign procedure, the main risk being the dangers inherent in sedating or anesthetizing the animal. Once the little dog was sedated, Dr. Mike would take a small, curving blade a bit like an apple-coring tool and lift out small samples of subsurface tissue. Then the samples would be put in formalin, a preservative, and sent to a lab.

(Dr. Mike had not done the procedure earlier because he was afraid that any sutures he made to Ari's flimsy, fragile skin after the biopsy wouldn't hold. It would almost be like trying to sew up a damp paper bag.)

Not long after the consultation with Dr. Angus, Dr. Mike sedated Ari to get him ready for the punch biopsy. When he laid him on the examining table, Ari had a cone-shaped oxygen mask over his face. He looked like an astronaut asleep on the moon. "We're going to find out what's wrong with your skin, once and for all!" Dr. Mike said affectionately to the dog. "OK, handsome?" Aristotle, already on the moon, did not respond. Once Dr. Mike had taken the biopsy sample and sutured up the wound—which did not burst after all—there was nothing to do but wait. "This is the hardest part," he said.

ANSWERS AT LAST

When the biopsy results came in, they were both surprising and unsurprising. Aristotle turned to out to have an extremely severe infestation of the *Demodex* mites that caused demodectic mange. Dr. Mike had been right after all. For some reason, the skin scrape just hadn't caught it the first time.

Did you know . . . dogs can get sunburned! Dogs with short hair are especially vulnerable. On hot summer days, protect their skin from the sun or keep them indoors.

But Aristotle's problems didn't end there. He also had a staphylococcal skin infection that was resistant to methicillin, the medication Dr. Mike had prescribed, so the vet decided to change his antibiotic. He also prescribed a new antiparasite medication, to mount a frontal attack on the infestation of mites.

And he decided to recommend that Aristotle be fostered—taken home with a staff member or volunteer, rather than kept at the sanctuary. Kristi Littrell, Adoption Coordinator, volunteered for this happy task.

Even though the housing for dogs at Best Friends is clean and spacious, and animals are regularly walked and provided with enriching playtime, fostering would help Ari on several levels, Dr. Mike said. For one thing, Kristi could more closely monitor the effects of the medication. Ari would get out of his cage at night and so have more room to run. That was crucial for a terrier, a breed with so much energy it seems capable of powering a small city. And of course, there was the ineffable healing power of a human touch.

"It feels good to help a dog like Aristotle, because he was such a miserable dog," Dr. Mike said. "It makes you feel better about what you can do. We don't give up hope on dogs here, and there's *reason* not to give up hope. Aristotle went from being a painful, shy, scab-covered mutt to a crazy, bouncy, not-itchy monkey. That's why we do this."

When Ari first arrived at Kristi's house, he seemed sad to leave his former home in the laundry room. He quietly explored the house, not making much noise at all. But he quickly grew used to Kristi's place and made himself at home. He barked first thing in the morning until the sun went down, telling Kristi his opinion on just about everything. His energy level began to soar, and his love of play returned with a vengeance. Kristi happily observed, "He loves life—everything is a party to him."

FINDING A HOME

Since his arrival, Aristotle had been a part of Best Friends' Guardian Angel program. The Guardian Angel website *(http://www.bestfriends.org/guardianangel)* features animals, often sick or injured, who need specialized care. It's a place where Best Friends members are able to connect with the cases that benefit the most from their support of the society. Each animal gets its own page, and staff update them with photos and progress journals that tell how things are going. (Even after animals are adopted, updates from their new families are posted as well.) Members are able to sponsor specific animals.

From his first days at Dogtown, Aristotle's page was populated with update after update, detailing the remarkable recovery of the feisty little dog. Members could see for themselves how far he had come—from the photos of his earliest days when his skin was so painful to view to the most recent photos where Aristotle sported a big smile and showed off his new, short coat of brown-and-white fur. Kristi knew that "there are people who, if there is not enough data on him weekly, they worry. They know he's in good hands, but they live for the updates."

Aristotle's biggest fans were inspired by his journey and sent many tokens of their appreciation to him. Gifts, notes, and well wishes poured in from all over the country, as well as Italy and France. Some people sent him food, stuffed toys, and little sweaters with "Ari" sewn onto the back. "Ari touches people," Kristi said. "He's a dog that would have been passed over in most places, but here he thrived. And he touches people's hearts."

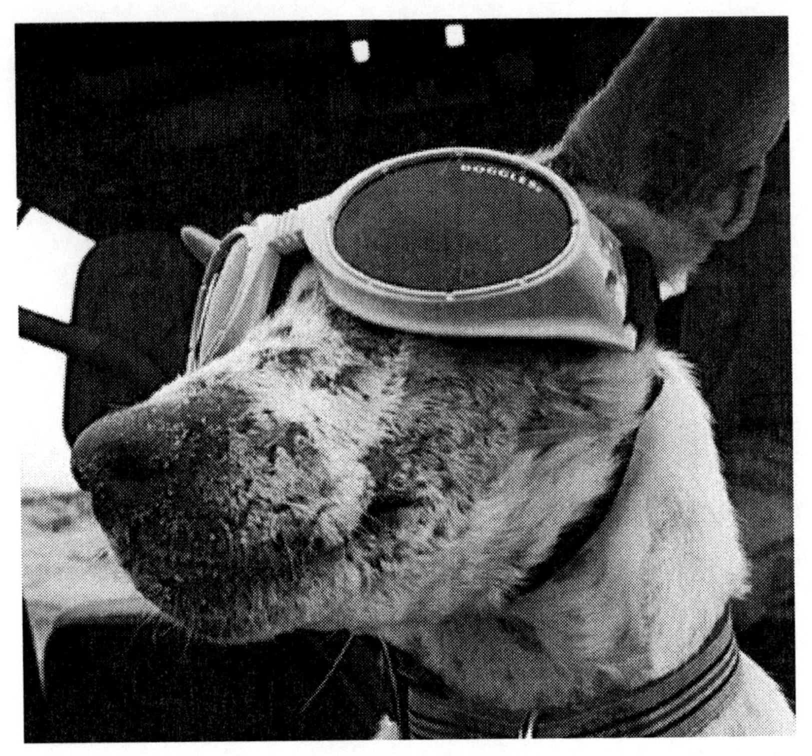

When he goes RVing with his new family, Aristotle must wear protective goggles to shield his eyes from wind and sand.

The Guardian Angel program is also where animals like Aristotle can find forever homes. Allowing members to follow a dog's progress gives them a way to connect and even fall in love. And that's just what happened with Aristotle. Applications poured in to adopt him, an embarrassment of riches.

Kristi and the Dogtown team had a tough choice to make and seriously considered every candidate. Because of Ari's skin, Dr. Mike thought a home in the northern states might be best to reduce his exposure to the sun. Kristi knew from fostering Ari that he could be needy, so the ideal home would have a caregiver present most of the time. And Ari's high energy level required an active home, preferably with other doggy siblings for Ari to play with.

From all the applications, a strong candidate began to emerge. They were a family from Oregon who had adopted from Best Friends before, a formerly shy spaniel mix named Jolene (also a featured dog on a *Dog-Town* episode). The parents, Susie and Phil, had eight dogs in all and were happy to invite Aristotle to be part of the family. They had a big house and yard, complete with a saltwater pool (which is perfect for a dog with sensitive skin, as the salinity helps soothe the skin). The more they considered it, the more Kristi and the Dogtown team were confident that this home was the place for Ari.

Once they learned their adoption application had been accepted, Susie and Phil brought their whole doggy menagerie to Kanab to introduce them to Aristotle before they took him home. Their first glimpse: a white flash of fur as Ari bounded out of Kristi's house and introduced himself. Next, after a series of introductions with them, he tore around the yard with Susie and Phil's dogs. The couple sat on Kristi's porch to watch the fun while talking with Kristi about Aristotle's needs. It was a perfect match for the little guy—he had plenty of playmates that shared his enthusiasm and energy, and he had two loving humans to look after him.

After taking him back home to Oregon, Susie and Phil still post frequent Ari updates to the Guardian Angel website. Phil reports that the dogs have worn a muddy "racetrack" around the pool in the backyard, where they love to run laps. But Ari is far and away the fastest of all the dogs. Ari's fans and friends can learn about his latest adventures—whether it's a trip to the groomer or his first camping trip, which included cruising in an RV over the dunes in a park (Ari needed to wear special goggles to protect his eyes, but he tolerated them like a trouper—once the vehicle started moving, he hardly noticed them).

Aristotle's high energy and joyful side might never have emerged if he hadn't made it to Dogtown. The sick, frightened dog he was upon arrival is now but a distant memory, thanks to the dedicated team of medical experts there. Today Aristotle is radiant and healthy—a bundle of fun covered with a healthy new coat of brown-and-white fur.

Shy Bingo dreaded meeting new people and other dogs.

Bingo: Opening a Shut-down Dog

Bingo was one of those swap-and-switch mutts whose body looked like it had been assembled by a committee of children; none of the pieces quite fit together. He looked to be mostly a yellowish Lab, possibly a shepherd mix, but with gangly legs suggesting there might have been some Great Dane or even greyhound in the mix. He had big, lugubrious brown eyes, sad as Eeyore's, as if to acknowledge how sorry he looked.

But when he first arrived at Dogtown on a dusty summer day in 2007, those sad, frightened brown eyes, peering out of a box, were the only part of Bingo that could be seen.

Dog trainer Ann Allums had just pulled into Dogtown after a long drive from southern California with a truck full of 14 new residents. She'd made the trip to pick up dogs from an animal sanctuary that was in the process of shutting down. The owners of the sanctuary, called Sage Canyon, had grown too elderly to continue running their operation and were sending their 14 toughest cases to Dogtown. Now several trainers, including Ann and Pat Whitacre, unloaded the carrying crates off the trucks, let the dogs out and leashed them up, and introduced them to their new runs. All the dogs seemed happy and grateful to be released after the long trip.

All of them, that is, except one—Bingo, who was so panic-stricken he refused to be coaxed out of his box. Immobilized by fear, he just cowered in the back of his crate. Finally, after Bingo had spurned all treats and entreaties, Ann and Pat had to pick the crate up and carry Bingo into his new home at Dogtown inside his crate.

"It was as if there were 14 crates, but only 13 dogs—the other dog was invisible," Pat said.

When he was finally enticed out of his crate, Bingo emerged with his body hunkered down low to the ground, tail tucked between his legs. He nervously investigated the run, taking in new sounds and smells, before he bolted for the "dogloo"—an igloo-shaped fiberglass doghouse—at the far end of the run. He crept into the sheltering darkness and "hid" there. Except for his face and eyes, Bingo's whole body was plainly visible through the doorway.

"I don't know what it was about Bingo, but as soon as I saw him, I said, 'I want to work with that dog,' " recalls Pat Whitacre. That was the day Bingo got lucky.

MORE THAN BASHFUL

But what drew Pat to this quivering, quaking creature? "I think I was partly fascinated by the 'mystery package'—the enigma in the box," he said later. "What kind of dog was in there? What was he like? What had happened to him to make him so fearful? And how dramatically could he be changed?"

Also, Pat has a special gift for working with shy dogs, perhaps because he shares a temperamental kinship with them. "I guess I'm kind of a hermit," he said. "I don't eat lunch in the staff room, don't socialize much after work. These dogs who just do not reach out and form bonds with people—I understand them.

"Shyness in dogs, especially dogs in shelters, is really an enormous problem," Pat said. "Dogs that don't express affection or come to people are much less likely to be adopted—so they're much more likely to be euthanized. People go to shelters looking for a pet, but they're also looking for an animal that returns a feeling of mutual warmth and connection, almost like a human friend. It's only natural."

So in the life-or-death cuteness contest of trying to win the affection of an adoptive family, shy dogs, like Bingo, are at a great disadvantage.

They're too scared to shine. They don't know how to display their sweetness. They could be warm-hearted, playful and obedient, an ideal pet, but without sufficient "people skills" to win a human heart, their lives are all too likely to be sad and short.

One other reason shy dogs in shelters tend to be overlooked, Pat said, is that people assume that they are always going to be like that. They think such a dog is incapable of change. They don't recognize that with love, patience, and training, even a dog like Bingo can be coaxed ever so slowly out of his box.

Beyond his life at Sage Canyon, little was certain about Bingo's background. But Pat didn't see this as a setback at all. As a general rule, Pat said, one of the biggest mistakes people make in trying to train a shy dog is concocting stories about why the dog is shy. They "get stuck" in a story, he said. The usual story line goes something like: This was a good dog until somebody did something bad to him, and he can be fixed if I am good to him.

"Actually, though," said Pat, "the dog's behavior often does not relate to what his experiences may have been as much as it does to his lack of experience. If a dog doesn't know what to make of people and has to learn, in fact, that these strange creatures are OK as a group, that's a slower process than trying to train a dog who had a traumatic experience that he's trying to recover from."

In general, Pat said, whenever a new dog comes into Dogtown, "it's best to assume that any or all of the information you have is inaccurate." Following this strategy allows a trainer to focus on the dog's current state of mind and on addressing existing behaviors in the moment. But wherever Bingo came from and whatever he'd been through, Pat's heart instinctively went out to him.

SHY-DOG SPECIALIST

Pat, with his gray beard, balding pate, and quiet, meditative manner, looks a bit like a garden statue of St. Francis. In a fast, frenetic civilization, Pat, 59, seems to be a man blessed with enormous reserves of stillness, patience, and calm. He's always loved animals. When he was younger he used to like to sit quietly in the woods and watch birds and squirrels, to see how they behaved. People said he had "a way with animals."

Of course, there are a few ordinary explanations for Pat's extraordinary patience: "When you grow up in a family with seven kids all waiting for the bathroom, you expect that good things will happen eventually, but maybe not as soon as you might've hoped," he said. "I've had to do a lot of waiting for things in my life, waiting to get results, for things to happen, so maybe my pace is a little slower than some folks might go."

On average, 10 million to 12 million animals are euthanized in shelters every year because they don't have a home.

He is also, he said only half-jokingly, "a person without direction." By which he means that he got to Best Friends not so much because he set out to arrive there, but because his life seems to have been guided by a series of "holy nudges" that caused all the puzzle pieces of his fate to fall into place. Pat likes to quote Fritz Perls, the father of Gestalt therapy, who used to say: "Don't push the river, the river will push you." What happens will get you there. That's how Pat Whitacre got pushed to Best Friends and Dogtown.

After getting his B.A. in psychology at the University of Kansas, Pat spent 30 years working as a mental health counselor. Most of his career was spent at Shawnee Mission Medical Center, an inpatient treatment center near Kansas City, working with the "chemically and persistently mentally ill"—human beings who found the world so terrifying and uncertain they had trouble coming out of their shell.

Later he earned a master's degree in biophysics and genetics at the University of Colorado. He was fascinated by the continuing nature versus nurture debate about what shapes animal (and human) behavior. His undergraduate degree trained him in the notion that it's primarily learned behavior, or "nurture"; his work in genetics, that it's mainly hardwired in the brain and body, or "nature." "It's helpful for me to see animal behavior from both sides," he said.

Pat, who isn't married and has no children, was at one point persuaded by a girlfriend to drive from Kansas all the way to Utah, to volunteer for a few days at the Best Friends sanctuary. When he got to Kanab, he realized that he had actually been there as a child: He remembered the statue of a cowboy on a white horse that stands on Main Street, framed against the red-rock canyon walls.

The river was pushing him.

When he got back home to Kansas City, he said, "I realized I had driven 2,400 miles to walk dogs, when there was a shelter about two miles from my house." So he started volunteering at the local shelter and gradually became "one of those hard-core volunteers who show up when the gates are locked for snow days." His background in mental health was helpful in working with dogs, because "even though the way human behavior develops and changes is not the same as what happens in dogs, the law is the same, or similar."

Two years later, he found out there was an opening for a dog trainer at Best Friends. His girlfriend encouraged him to apply. "I said, 'I'm not a dog trainer, I'm a mental health counselor who walks dogs!' " But he applied anyhow, arranging to come to Kanab for the two-week trial stay required of most applicants. He arranged to get two weeks off from work at the hospital, and was almost ready to go when he realized he didn't have a travel alarm clock, which he'd need for the trip. That same day, a free travel alarm clock came in the mail, as part of a fund-raising appeal from the Humane Society of the United States.

It was the river again, pushing him in the direction he needed to go. Now, four years later, Pat Whitacre's office is a ten-by-sixteen-foot room at the Dogtown sanctuary.

BETTER SAFE THAN SORRY

When he first got to Best Friends, Bingo's whole approach to life was: "Better safe than sorry." Pat said, "There's no sense to go boldly charging into things you don't know about, or that you aren't convinced are going to be safe for you. So it's a fairly wise choice for a dog like him. His challenges are to find out that the world is, in fact, OK."

Showing signs of nervousness Bingo, right, tucks his tail and turns away from his trainer Pat Whitacre and Peanut.

Shyness in dogs can express itself in different ways, Pat added. Some dogs deal with their shyness by fleeing scary situations as fast as they can, running away, hiding, avoiding. Other shy dogs express their feelings with what he calls the "puffer fish" approach. They try to puff themselves up as big as possible, to become frightening and aggressive. Make the enemy flee first; don't let anything get close enough to hurt you. If anything gets too close, bite—a damning behavior in a world run by humans, because it's inclined to get the dog killed.

Bingo had chosen the first shy-dog approach—luckily, the easiest one to work with. When Pat approached him, he would slink away and try to hide, tail tucked between his legs, hunkered down so low it was as if he were trying to melt into the ground. He didn't show any biting behavior. His whole demeanor was passive—he seldom even avoided people with any real force.

At first Bingo did not readily respond to Pat or anyone else. There was no evidence that the dog had ever developed a real relationship with a person. Instead of responding to humans, Bingo reacted with a behavior Pat called "shutting down." Bingo would try to turn into an inanimate object, like a rock or a plant, in the hopes that he would go unnoticed. He would lie down or hide in his doghouse, holding very still and staying quiet. Pat describes it as "almost like hitting the switch in a power plant and the whole thing just shuts down."

FINDING BINGO'S REWARD

But Pat saw Bingo's shutting down as an advantage for a trainer. His strategy of keeping still meant that he wouldn't bite and he wouldn't run away, so it was easier for Pat to get close to him, to touch him, eventually to slip a leash over his head—to take the first tentative steps toward molding his behavior. And with shy dogs, every step may seem small, but even these little things can be considered big successes. Pat would build on these moments to try to bring Bingo out of his shell even more.

Bingo's shyness was not as extreme as that of some dogs Pat had seen. He was easier to work with than a dog that isn't comfortable with people being close at all, or with people touching it, because trying to teach such a dog to sit is "a hopeless and futile task," Pat said. "You're talking to him—he hates that. You're close to him—he hates that. You offer him a treat—he's too nervous to take it, so you can't reward the behavior. The only real reward with some extremely shy dogs is to go away and leave him alone—which is the technique you sometimes have to use."

Bingo, by contrast, was at least approachable, which meant he was teachable.

Even so, Pat knew there were special difficulties in working with shy dogs. The more behaviors a dog offers to its trainer, the more relaxed the dog is, the easier it is to shape that behavior, he said. A confident, expressive, outgoing dog, one that offers many behaviors, even if they are undesirable behaviors, is easier to train than a shy, shut-down dog. If a shy dog is doing nothing, it's very hard to reward the behaviors you want to see, especially if he or she is too nervous to accept the rewards you're trying to offer. That was the challenge with Bingo. He was so shy, so shut down, so trapped in the box of his own fear, that he had trouble making progress.

Still, one of the great rewards of working with shy dogs, Pat said, is that although it's "real easy to just see the problem behavior and not realize there's more to the dog than that," there's enormous soul satisfaction in those moments when an animal begins to show parts of himself that may at first have escaped notice. Who knew what winsome and endearing qualities Bingo might be hiding inside the box of his own shyness? Pat knew that the key to the whole process would be finding out what Bingo liked and then using it to reward him.

Generally most dogs come to Best Friends by car, truck, or commercial flight. In emergency rescue situations, Dogtown has access to a propeller plane that transports dogs to the sanctuary.

After working with Bingo very gently, Pat took him out for his first walk. When the leash slipped over Bingo's head, he tolerated it in a passive way, like many shut-down dogs do. He did not fight the leash or invite it; he just ignored it. Unlike many dogs, who get excited at the first sign of a walk, Bingo merely seemed resigned, neither excited nor resistant. Bingo walked alongside Pat toward the door of the enclosure, never looking at Pat the entire time, even after they exited the pen and began the walk. Many shy dogs will walk behind the trainer, to keep him or her in constant view, or if they're ahead, they'll constantly check back to make sure the trainer is not gaining on them. But Bingo walked beside Pat, sometimes even bumping into him with his body. Although Bingo was tolerating the walk, he was not relaxed; his tail was down, his body slunk down, low to the ground, and he would occasionally startle at noises. But although Bingo didn't seem to enjoy the walk, he didn't hate it either, and overall did "pretty good" on his first walk, Pat said.

The second time Pat entered Bingo's enclosure to take him for a walk, Bingo came up to the gate by himself, a sure sign of progress. Despite his resigned attitude, Bingo had enjoyed his first walk and was looking forward to another one. This was a big moment for Pat—he had found a reward for

Bingo, something to help encourage new, good behaviors. So Bingo had given Pat more useful information about himself, demonstrating that he wasn't so shut down that he couldn't take pleasure in some things.

LIFE WITH PAT

Pat decided to take Bingo home as part of Dogtown's foster program, which allows caregivers and volunteers to take an animal home to help it grow accustomed to living in a human household, and also to assess its progress. Pat wanted to take Bingo home for a couple of reasons. For one, he just really liked the guy. In fact, he was thinking of "just being really selfish and keeping him." For another, Dogtown is a busy, bustling place, full of animals and people, which could be overwhelming to a dog who could tolerate only a little bit of human interaction. Pat felt he could make quicker progress with him in a smaller, quieter atmosphere, where there were only one dog, Rolly, and two cats.

On the day of the move, Bingo cowered in the back of Pat's truck during the short ride to his house, a double-wide trailer on the edge of Utah's high desert. He had to be coaxed out of the truck, eventually slinking out with his tail between his legs. He was panting constantly, a sure sign of stress. Bingo was so nervous he refused to take treats from Pat's hand. And though Rolly came bounding up and greeted Pat with a direct gaze and lots of kisses, Bingo avoided Pat's gaze completely.

Once Pat got him inside the house, Bingo tried to bury himself under a bed or a nightstand or a chair, as if, if he couldn't see anyone, no one could see him. It was laughable to see this 70-pound dog with his head buried under a tiny nightstand, rump in the air, like Big Bird, thinking he couldn't be seen.

But Bingo began to fit in to Pat's household. He was perfectly house-broken. He also got along fine with Rolly and the cats, Bright Eyes and Buffy. Bingo wasn't inclined to create problems in a household, at least Pat's household, a trait that would help him when he was ready to be adopted.

Pat began taking Bingo from his home to the office every day, to expose him to two different environments, one safe and quiet, the other a bit busier and more inclined to produce the unexpected. Gradually, a bit of daylight began to shine under the lid of his box. He'd emerge from his crate with his tail wagging when it was time for a potty break. He'd spend time out in the room when Pat was gone and other people were there. But at the slightest surprise—an unfamiliar animal, a sudden sound—Bingo would bolt back to the safety of his crate.

Pat came to be able to read Bingo's emotional states by the way he held his body. Usually he held it low to the ground, slinking along as if he were very scared, with his tail dropped, his ears down, and his neck low and extended straight out in front. But when he began to relax, there was, Pat says, "an altitude change." Bingo's whole body lifted up and got bigger. It was as if he allowed his body to become as big as it actually was, instead of trying to shrink it down to the size of a Chihuahua. Allowing himself to get big also meant that he was not afraid to show himself, to be seen. His ears and tail lifted—at least a little bit—and he began looking around as if he were curious about the world.

One deficit Pat noticed was that Bingo did not solicit play from another dog, and no other dog solicited play from him. Yet anyone who has spent more than 30 seconds with a happy dog knows how important such play is. One day Pat took Bingo out with another dog and encouraged them to play in a shallow pond, but Bingo just high-stepped daintily around it, not wanting to get wet and seeming to take no pleasure in any of it.

"Asking if play is important to dogs is kind of like asking if it's important for people to enjoy themselves in their life, or if they can just exist, kind of taking care of business," Pat said. "Play and fun are all important parts of being alive, part of making the trip enjoyable. If we get a dog that just exists, that's just eating and sleeping, we haven't done much for him. So it's an important step to see when they're starting to be able to loosen up enough to have some fun."

Pat decided to try seeing if he could get another dog to induce Bingo to play. When a dog invites another dog to play, it will bow down low on its front paws, raise its rump in the air, and wag its tail. "It's as if the dog

Pat Whitacre tempts Bingo with a treat to see if this reward might motivate Bingo to overcome his shyness.

is saying, 'Whatever I'm going to do next isn't serious, so don't take this wrong—let's just go have some fun,' " Pat explained. "That's important because playing often involves a lot of behaviors like chasing, stalking, mounting, play-fighting—things that could easily be misinterpreted by the other dog if they didn't understand this was play." The initial bow lets everyone in on the fun.

Pat took Bingo and a frisky little dog named Joey to Tara's Run, an enclosed "playground" at Dogtown filled with obstacles—jumps, a tunnel, a seesaw. But though Joey bounced around madly, delighted to be off-leash in a world filled with fun, Bingo just seemed overwhelmed. He reverted back to hiding and avoided both Pat and Joey. The experience was just too much for him; Bingo still wasn't comfortable enough to play.

Every day, Pat continued to work with Bingo. He took him for walks at home around his neighborhood, which Bingo really seemed to enjoy, particularly the longer ones where they ventured far from the house. But even on these walks, any kind of distraction—kids on bikes, men hammering on a roof—could spook Bingo. It seemed to take very little to disrupt the fragile inner calm of his life.

But day by day, walk by walk, Bingo began to show signs of improvement. His bond with Pat grew stronger, which helped him feel safe enough to relax, When Pat had visitors to his house, Bingo would sometimes prowl around a little instead of hiding. Sometimes Bingo would come to the door when Pat came home and do a big doggy stretch, then roll over on the floor on his back. And when Bingo woke up in the morning he'd jump on the floor and roll onto his back as if he were giving himself a delicious, lingering back scratch.

These gains were small—emotional growth "on a shy-dog scale," as Pat calls it. But, he said, "It's always a mistake to shortchange an animal by saying that you know how far they can go. Because you just don't know how far they can go."

RETURN TO TARA'S RUN

As Bingo grew to trust Pat more, Pat decided it to try playtime again. He wanted to go back to Tara's Run with Bingo and bring along three other dogs, too. Instead of inviting Bingo to play, these three dogs would show him how to have fun. Pat hoped that the combination of his reassuring presence and the example of the other dogs could coax Bingo into having a good time, a sure sign of progress.

Bingo's playmates—Rolly, Sylvie, and Sarge—were ready for fun as soon as Pat took off their leashes. When Bingo's leash came off, he started to look for a place to hide until Pat broke out the treats. The three dogs eagerly surrounded Pat in anticipation of a snack. Bingo started to move away from the group, but then, for the first time, he stopped. Pat praised him immediately, "Good boy, Bingo. Treat, Bingo!" and handed him a morsel of chicken. Bingo, a dog who had once been so shut down that he shunned treats, accepted the reward.

Building on this first success, Pat began jogging around the playground. He knew that Bingo looked to him for security and was likely to stick close to him. As Pat jogged, he happily called to the dogs

"Good job, Sarge! Here we go, Sylvie! C'mon Rolly! Let's go, Bingo!" The three other dogs frisked happily around Pat; at first, Bingo seemed nervous and ran ahead of the group. But as the exercise continued, Bingo began to loosen up and joined in with the rest of the dogs. As he trotted alongside Pat, his body language began to change. Slowly, his tail began to lift higher and higher. He no longer held his body low to the ground, tense and apprehensive. Now, his posture straightened and he bounced happily as he did laps around the playground. Pat began to lead Bingo over some jumps and through a tunnel, rewarding Bingo for each performance with treats and praise. And Bingo responded with a more relaxed demeanor. Pat ruled the outing a success: Bingo was having fun.

Dogs have been important to humans throughout history. From assisting hunters to guarding against predators, dogs are among the first and most valued domesticated animals.

Pat knew that with time, more of Bingo's behavior would change. He could tell that Bingo was "starting to enjoy being a dog." He knew Bingo would frolic. He'd put his ears up and wag his tail. He'd get excited about going someplace. Little by little, the lid of Bingo's box was lifting. The daylight of life—with all its richness and danger and delight—was flooding in.

It was impossible to say what was going on inside Bingo, but it did seem as though his life were becoming sweeter, less scary, more . . . joyful.

And for Pat Whitacre, that was the great reward.

A Is for Atticus

Pat Whitacre, Certified Pet Dog Trainer

I suppose, when people examine their life for pivotal moments or characters, by definition they find themselves looking at beginnings and endings. Armed with that perspective I really must start with ***A*** and tell the story of Atticus. After a visit to Best Friends in 2003 helped me discover a way to channel my love of animals into constructive action on their behalf, I began to volunteer at a local shelter in my hometown. I knew a thing or two about behavior, I told myself. I would help the dogs with "issues" become more adoptable.

The shelter staff seemed appreciative—or at least, polite—in their amusement. They were quick to offer suggestions of some "issues" dogs I might work with. After all, many of the more "challenging" dogs exhausted volunteers in short order. They often were left till last if they got any interaction at all, which tended to make the situation worse. Who was at the top of the list? A beautiful 80-pound German short-haired retriever named Atticus. He was fairly young, maybe three or four years old, healthy, and ***energetic.*** Atticus did not have a mean bone in his body, but anyone who went into his run or took him out for a walk emerged feeling like they had gone three rounds with Evander Holyfield. He would jump, push, pull, shove, ram, drag, wag, or otherwise abuse your body until you escaped his enthusiasm.

In truth, Atticus had many qualities that made him a good candidate for training. His friendly nature minimized any risk of aggression. His high activity level produced a lot of different behaviors that I could try to reinforce. The only real question was whether I could figure out what rewards he would respond to. As I got to know Atticus better, the answer revealed itself: Just getting to go and ***do*** things was reward enough for him.

Atticus made quite respectable progress. He began keeping his feet on the floor when I entered the run. He would sit to be leashed and wait for a signal to go through the gate. He spent less time dragging me on the leash, partly because of his training, and partly because we jogged for three miles each day to start our outing. After all, sitting in his run 23 hours a day provided little outlet for his enormous energy. Having bragged about his progress, I should mention that these improvements were largely confined to his interactions with me. He had not yet generalized his new behaviors to other people. They did not behave like me, so he assumed the old way of doing things was the way they wanted him to behave. Just because he had to put up with one person who behaved strangely was no reason to think everyone else had lost their mind at the same time.

All this so far is prologue. The important part of this story is that one winter day a man and his ten-year-old son came to the shelter to adopt a dog. This shelter always had lots of nice, easy dogs from whom to choose. So, of course, they went right to Atticus's gate and fell in love at first sight. The staff explained his history and foibles to the gentleman as diplomatically as they could. Atticus was not considered the best choice for a family with children under 16, since adult volunteers found him overwhelming. But, as the saying goes, "the heart wants what the heart wants," and the man and his son were not easily deterred. In keeping with the time-honored technique of dealing with an impasse by handing it off to someone else, the staff politely introduced the man and boy to me. They explained that I had been working with Atticus, that I knew him better than anyone else at this time, and that I would be happy to bring him out so that they could meet him. I sensed impeding disaster. As mentioned before, Atticus did not expect to use his new behaviors with "normal" people, so his high-energy antics could be on full display. I suspected that the dog weighed as much as the boy. Did I mention that it was winter? I thought so. The play yard we used for introductions was covered in snow.

The young boy wanted to walk Atticus first. I wanted Dad to try but I was outvoted two to one. As Dad and I watched, the boy proudly took the leash. Almost immediately Atticus took off at a full run, dragging the boy behind him. I glanced at Dad to catch his reaction. To my surprise, he was smiling! I turned back around, and instead of seeing the poor lad facedown in the snow I saw the youngster with both legs locked stiff, remaining upright, skiing behind Atticus as the pair zigzagged back and forth across the lawn! He was having the time of his life. Dad looked at me and said, "He's perfect. Can we have him?"

Therein lies the beautiful truth about dogs. One man's problem dog is another man's joy. With the best intentions, we at Best Friends try to help dogs become more adoptable. We encourage behavior that would seem to open more doors to potential homes. But, in the end, the more pertinent goal is to find the right match between home and dog. I still believe there is value in helping dogs change some of their behaviors, but I realize that the value inherent in their individuality should not be underestimated. In this case, the "magic" happened not because of my efforts to shape Atticus into some generic adoptable dog, but precisely because of the unique qualities that made him Atticus.

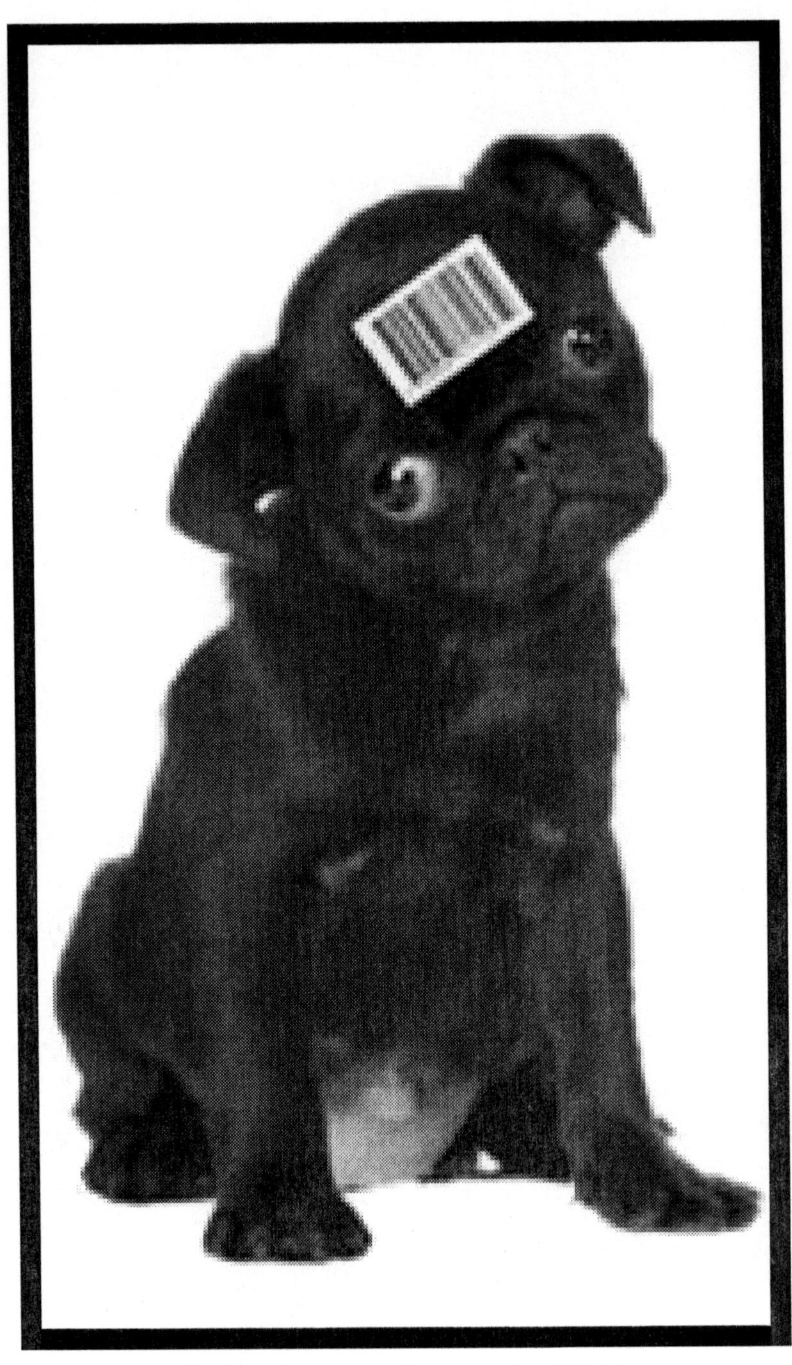

A cute pug is the face of Best Friends' Puppies Aren't Products campaign.

Parker & Mei Mei: The Problems with the Puppies in the Window

It's hard to imagine anything cuter than puppies in a pet store. Those precious faces! Those adorable eyes! Those bumpety-bumping tails!

But all too often, love-struck buyers don't ask or even wonder where the puppies actually came from. They may envision a peaceable countryside filled with gamboling, happy pups. But unfortunately, the truth about the puppies in the window is often almost incomprehensibly grim.

Many dogs sold in pet stores come from puppy mills, large-scale commercial operations that breed dogs for profit. (Another term for puppy mill is "commercial breeder.") The Humane Society of the United States defines a puppy mill as a mass-breeding operation where dogs often live in overcrowded, unsanitary conditions. The dogs are bred for years solely to provide puppies for the pet trade. "A puppy mill is like factory farming for dogs—treating them almost like livestock," said Best Friends Dog Care Manager Michelle Besmehn. "These breeders are breeding dogs over and over again, often without considering the animals' health at all."

In many operations, the adult breeding dogs are the most mistreated animals of all, since the public never sees them. They can be subjected to terrible conditions, spending 24 hours a day, seven days a week in a cage, often with little social interaction with humans. They eat, sleep, and eliminate all in the same wire cage. Female dogs are often bred every time they come into heat, so they spend most of their lives either pregnant or

When Parker came to Dogtown, his teeth and gums were in such bad shape that just chewing his food caused great pain.

nursing. A single female may bear 60 pups; when she's too old to bear any more, she's simply killed.

"A responsible breeder wouldn't continually breed their dogs over and over again, basically breeding them until they wear out," said Michelle. "They would only have one or two litters and make sure the breeding pairs are healthy, that they don't have any behavior or medical issues. But that's not the case in a puppy mill."

Large-scale breeders like this are legal in all 50 states, but they're poorly regulated. Today the Humane Society of the United States estimates that there are approximately 10,000 puppy mills nationwide and only 100 federal inspectors to monitor them. Regulations about the number of dogs and quality of conditions tend to be fairly vague. These facilities got their start after World War II when the U.S. Department of Agriculture encouraged struggling farmers to breed puppies to make some extra money. Today, besides supplying pet shops, these kinds of operations often supply Internet merchants, feeding the insatiable demand for a cute face and a wagging tail, no matter what the origin.

At the pet store, a puppy may sell for a few hundred dollars up to over a thousand—a premium price for what the new owner hopes is a heartwarming bundle of joy that is a good example of the breed and also reasonably healthy and well adjusted. "But what often happens is that people get the puppies home and either they have medical issues or behavior issues that they didn't expect," Michelle said. "Or the puppies grow up and have some issue that someone doesn't know how to deal with, and they end up in shelters. So it ends up adding to the overpopulation problem."

It is from two large-scale operations that Best Friends rescued two true survivors, a dachshund named Parker and a Chihuahua called Mei Mei. The two dogs came from different locations, but the medical and social issues they faced are all too common in dogs that come from similar situations. Luckily for them, they found their way to Best Friends, where the dogs' strong spirits and giant hearts would be nurtured by the caring staff at Dogtown.

PARKER: THE DOG FROM WHISPERING OAKS

The website was cheerful and innocuous, featuring a photo of a dachshund wearing a baseball cap and sunglasses. "Specializing in miniature Dachshunds!" read the text, which advertised the services of Whispering Oaks Kennels, in Parkersburg, West Virginia. "Happy, healthy puppies are our priority," it went on. The site mentioned, in passing, that although visitors were welcome, they were not allowed into the kennel area where adult breeding dogs were kept. It ended with a cheery adieu: "Thanks for Visiting Our Site and God Bless!"

It was easy to imagine dachshund puppies frolicking on a golden hill, with mighty oaks whispering overhead, but the circumstances at the kennels were far different. Nearly a thousand dogs were confined in an assortment of small cages. In some cases, four or five dogs lived in one two-foot-by-three-square-foot cage. When the animals were first let out of their cages, many of them stumbled as their feet touched grass, tile, or carpet—perhaps due to muscle weakness or to the unfamiliarity of the surface beneath their feet.

The local sheriff's office began investigating Whispering Oaks in summer 2008 after a former employee complained about waste disposal practices there. The owner was not charged with animal cruelty or neglect; she agreed to downsize the operation and voluntarily surrendered more than 900 dogs, who she insisted had received regular veterinary visits, had been well cared for, and had never been mistreated.

A DACHSHUND WITH TERRIBLE TEETH

In the days that followed the dogs' surrender, a large number of rescue organizations, including Best Friends, the Humane Society of the United Sates, the Humane Society of Parkersburg, and other rescue groups, worked together to find new homes for them. Most of the surrendered dogs from Whispering Oaks were adults—breeding animals that would simply have been put down when their fertility began to wane. The 928 dogs—mostly Chihuahuas, dachshunds and poodles—were relocated to a local warehouse, from which volunteers would determine where they would go. It was a noisy, hectic scene, with volunteers bustling around, cleaning cages, doing medical exams, filling out paperwork, and generally getting the dogs ready to go out to rescue groups and a new, better life.

As part of her job as Dog Care Manager at Best Friends, it was Michelle's unenviable task to choose which dogs to take back to Dogtown. "It's pretty overwhelming—how do you choose?" she said. Michelle was particularly interested in the animals that might have the most trouble finding homes: the ones in needed of medical care and older dogs. Best Friends' excellent medical clinics are something few other shelters can afford to have.

"I told the volunteers what kind of dogs we were looking for and asked them, 'Are there other dogs you'd like me to look at?' " she recalled. "That's when they showed me the dachshund with the terrible teeth."

DENTAL DISASTER

From the rear, the little long-haired dachshund didn't look special: a reddish brown, hot dog body, four legs, and a tail. But from the front, all you could see were teeth, so long and so misshapen they resembled the plastic fangs kids wear at Halloween. Even when his mouth was closed, they poked out and distorted his face into a crooked grin.

The dog's misshapen teeth might have been the first thing people noticed, but the second was definitely his breath. It was appalling and an indication of possible infection. Michelle also noticed that he seemed to eat very gingerly, as if his mouth hurt.

Bad teeth are a problem typical of many breeding dogs in puppy mills. Dental work is generally considered an unnecessary frill and a potential threat to the bottom line. Problems of neglect, like bad teeth, matted fur, overgrown nails, and eye and ear infections, are allowed to fester, causing pain and greater suffering for the dogs.

The midwestern United States has the highest concentration of puppy mills. It is thought that Missouri has the most, followed by Oklahoma, Iowa, and Arkansas.

This toothy dachshund had become a favorite of the people working with the Whispering Oaks dogs. When people came up to his cage, he would timidly approach them with this strange, crooked grin on his face. The dog's quirky expressions charmed everyone who saw him, and their positive reactions encouraged his advances.

But even though the little dog welcomed the attention, he wasn't quite sure what to do with it. When visitors reached into the cage to touch him, his wiry body shied away from contact, showing how uncomfortable he was about being petted. Michelle's guess was that he was not afraid of people, but he had never been picked up and cuddled very much. The paperwork on his cage explained that he had been found living in a tiny rabbit hutch with three other dogs, a space he had probably rarely left since puppyhood. For shelter, he and the other dogs shared a small, bare plywood box with no bedding material at all. His tiny, limited life most likely never included a warm bed.

When Michelle first came across the wee dog with that odd, endearing snaggletoothed face, she was charmed by his smile too. This small, middle-aged dog with medical issues was perfect for Dogtown, and Michelle decided to bring him back home. She named him Parker.

NEW DIGS AT DOGTOWN

After Parker's carrying crate had been loaded into a van for the 40-hour drive from West Virginia all the way back to his new home in Utah, Michelle peeked into his crate and noticed that Parker had that odd, weirdly endearing jack-o'-lantern grin on his face. Maybe he knew that this extended confinement in a tiny cage—the sort of place he had spent much of his life—would soon be his last.

Michelle knew that Parker's mouth would need to be examined very quickly once he arrived. It wasn't just that he had crooked teeth and bad breath, but also that he was clearly physically uncomfortable. Chewing seemed to be so painful that volunteers began soaking his food to soften it—but even then, it still hurt. He seemed to bite things gingerly, and if they were too hard or too large, he would simply nudge them aside in his bowl. Signs of infection were also worrisome. Just by peeling back Parker's lips, you could see what Michelle described as "green goo" on his teeth and gums. What it was, she shuddered to think.

When Dr. Mike Dix, Dogtown's head veterinarian, looked into Parker's mouth, he was taken aback. "I've seen some pretty bad teeth in my day, but Parker certainly ranks up at the top," he said. The more he looked, the worse it got. Dr. Mike determined that several of Parker's teeth were going to have to be pulled; some were so loose that he probably could have pulled them out with his bare hands.

Parker's mangled teeth could have been the result of bad breeding—a genetic defect that was allowed to continue down through the generations. Dr. Mike felt disgusted that the puppy mill was probably using Parker as a "stud dog" for breeding, even though his poor teeth made him an obviously bad candidate. "But it doesn't surprise me, because that's what a puppy mill does—they breed anything they can breed."

Parker's shaggy red coat became soft, shiny, and lush after a good grooming at Dogtown.

The infection in Parker's mouth also concerned Dr. Mike; it was so virulent that it could become life threatening if it spread to his bloodstream and then to his heart and other organs. Left untreated, Dr. Mike said, it could kill him within a year or two.

"He's a high-priority dental—let's move him to the top of the list," he said to Michelle, explaining that at least several of his teeth would need to be extracted.

A few days later, after his initial exam, Parker seemed to be making a quick and successful adjustment to his new spacious digs at Dogtown. His shaggy red coat needed a good grooming, a process that Parker really seemed to enjoy. After a shampoo, a blow-dry, and brushing, his reddish brown coat gleamed. With a shiny coat and bright eyes, he prepared for his surgery with Dr. Mike.

Before his operation, Parker was prepped for surgery and then sedated. Pulling teeth is traumatic for an animal, Dr. Mike explained, so he wanted to limit the number that were extracted. The canines were of special concern; they were such large teeth that the gum would have to be sutured to avoid creating a fistula, or abnormal opening between Parker's oral and nasal cavities.

Dr. Mike called in Steve Lund, a vet tech who specialized in dental examinations, to help with the surgery. When Steve took a look into Parker's mouth, he gasped a little. "It's pretty bad . . . wow," he said. "And his breath . . . whew!"

Then he pulled out one tooth with his fingers. This was a disquieting discovery. It appeared that the infection had spread to Parker's jawbone. X-rays revealed that nearly all of the little dog's teeth were rotten. Steve and Dr. Mike decided to remove all of Parker's teeth. Only time would tell if this drastic surgery and a regimen of antibiotics would prevent the infection from spreading.

A change in eating behavior can indicate mouth pain in a dog. Eating slowly, dropping food, or tilting the head while chewing, can all be signs of discomfort.

PARKER'S RECOVERY

After the surgery, Michelle could tell that Parker's mouth was extremely sore. If before he had taken bites gingerly, now he barely took any bites at all. He seemed sluggish and dispirited, almost as if eating were too much effort to bear. He did not seem to be recovering as quickly as the staff had hoped.

Michelle decided to see if she could get him to eat a little bit. He looked up at her with his sorrowful, slightly hurt brown eyes, and despite his discomfort, gave her a no-longer-crooked grin. She noticed that there was some canned chicken and some kibble in his cage already, but that it was uneaten. When she tried to feed him, it became apparent that he was having a difficult time simply getting the food into his mouth. So she piled his food up into a little mound, to make it easier for him to grab with his toothless jaws. He began wolfing down his lunch hungrily, so

Michelle made another chicken-kibble mound in a different spot in his cage. Parker loved it. He scurried back and forth between the two piles of food, ecstatic to have two meals to fill his hungry belly.

Dr. Mike kept checking on Parker regularly in the days after the surgery, to make sure his mouth was healing properly. Three weeks later, Parker was making a remarkable recovery, with the oral infection under control. He was also learning how to eat without teeth, and his flagging spirits seemed to have revived. He eagerly greeted visitors, padding up to the edge of his cage and looking up beseechingly, holding up his wet brown nose as if offering it to be petted. Parker was on the mend.

It was no surprise to Michelle. "Dogs are amazingly resilient when it comes to things like losing all their teeth," she said. "I've always admired that in dogs. . . . My guess is he's going to move on, figure out how to eat with no teeth, and he'll be fine.

"People tend to hold onto things more, psychologically. If you lose all of your teeth, you're going to think, Well, what's my mouth going to look like? How will this change my self-image? and it can be quite traumatic. Dogs seem to be able to move forward more quickly than people."

PARKER'S NEW HOME

Now that Parker was on the road to recovery, placing him in a loving home became Dogtown's next goal. He needed to find a place with a family who would cherish his endearing oddities, including his total lack of teeth. His first move was into a foster home with Juliette Watt, a Volunteer Coordinator at Best Friends. Juliette knew that a dog like Parker could make huge strides in his recovery just by living in a home environment. The first night there, it was as if Parker could not believe his luck. He ran from room to room, exploring the vastness of Juliette's house.

But when he took his first steps outside, the wide-open space was like heaven to the dachshund. Parker reveled in the myriad smells in his nose and the new textures under his paws. During his first week, Parker found a favorite place on the backyard deck. With his ears erect and his eyes alert, Parker perched up high and watched the world go by. Within weeks, Parker's confidence grew and grew. He quite literally found his voice and began barking for the first time since he had left Whispering Oaks. His mouth was healing nicely from the surgery, making it easier for him to eat. He made friends with Juliette's dogs, especially her dachshund Rosie. Things were looking up for Parker.

From the beginning, Parker's story had been featured on Best Friends' Guardian Angel website *(http://www.bestfriends.org/guardianangel),* a place dedicated to showcasing the stories of the shelter's special-needs animals. Most of these animals have conditions requiring a lot of medical care, and these pages offer a place where viewers can monitor their progress. Best Friends' staff update each animal's progress journal with news and events. Through the Guardian Angel program, sponsors can donate money toward the care of these animals and people can apply to adopt them. It was through this amazing program that Parker found his forever home.

After the story of Parker's rescue and surgery was featured on the Guardian Angel site, it only took a short time before a potential adopter appeared. Parker's moving tale charmed a woman named Becky from St. Cloud, Florida. Becky had six rescue dogs at home, two of which were dachshunds. It would be an ideal environment for Parker, one in which he could learn to socialize with other dogs as well as with his new people. Becky and her family were well aware of Parker's background and felt up to the challenge of giving this boy a good life.

Becky traveled all the way to Dogtown from Florida to meet Parker personally and to bring him to his new home. When Michelle brought Parker out from behind the counter at the welcome desk, she said to Becky, "Would you like to hold him?"

"I'd love to!" she answered.

"OK, Parker, be a good boy," Michelle said, handing him over. "I know you're going to have a good life!"

Parker gave Michelle one of his sly—now toothless—old Parker grins. Then Becky took him and held him tight, and Parker didn't seem to mind one bit.

Mei Mei's big personality charmed the Dogtown managers who traveled to Los Angeles on a puppy mill rescue mission.

MEI MEI: BIG-EYED LITTLE SISTER

In puppy mills, the breeding dogs often have medical problems, but there is another huge, mathematical problem: Puppy mills produce an estimated four million dogs for sale each year in the United States. At the same time, there is an enormous overpopulation problem, with an estimated three to four million homeless dogs and cats killed annually, according to the Humane Society of the United States.

"It's a national travesty that the puppy mill industry is allowed to flood the market every year with so many dogs being killed each year in animal shelters," said Julie Castle, Best Friends' Director of Community

Programs. For the average animal lover, the solution can be summed up by the slogan of one recent campaign: Puppies Aren't Products. This campaign has a logo of a soulful, sad-eyed pug puppy with a bar code across its forehead.

Because the problem is so widespread, Best Friends has developed several programs within its Puppies Aren't Products campaign to address the whole issue at several different levels, from the actual facilities themselves to the pet stores that are knowingly buying and then reselling these dogs. The campaign also seeks to educate buyers about the problems. The goal is to help shut down more puppy mills and encourage people to adopt homeless dogs from a shelter.

In 2008, Best Friends brought its Puppies Aren't Products campaign to Los Angeles to fight breeding in and sale from puppy mills and their importation in the Los Angeles area. After an undercover investigation by a Los Angeles-based animal rights group, Last Chance for Animals (LCA), in conjunction with the local CBS-TV news station, a Los Angeles County puppy mill was found to be keeping 402 dogs in inhumane conditions. Among other outlets, the operation had been supplying puppies to a pet store in Beverly Hills called Posh Puppy, where they were selling for as much as $4,000 apiece.

Since the kennel was only licensed to keep 100 dogs, it was ordered to release the remaining 300. Approximately 200 were sold or passed along to other breeders; 50 were dropped off at the Lancaster Shelter, where they were quickly placed by local rescue groups; and another 40 were rescued by a joint LCA/ Best Friends operation. Michelle traveled 450 miles from Utah to L.A. to see which of the dogs most needed the special care that Best Friends specializes in providing.

When she arrived, Michelle found the caged dogs had been numbered for identification purposes, but they had no names. She was looking primarily for elderly dogs that might have medical problems or might be unattractive in other ways that made them difficult to adopt. It was as if she were answering a reverse personals ad: "Seeking someone old, sick, and ugly."

Michelle and her assistants chose six small dogs and were about to call it a day when she sat down with a Chihuahua who was not old, or sick, or ugly. The little short-haired dog, with fur the color of banana pudding, had large, round, expressive eyes, with a black muzzle and a pronounced underbite. Her upper lip kept getting snagged on her teeth, exposing a single, comical fang, as if she were attempting to look fierce and failing. She was wearing a pink collar bearing the number 15.

The little dog started prancing daintily around the enclosed side yard at the foster facility where the dogs were being kept. Her upturned, curlicue tail, almost like a little pig's, bounced and jiggled as she frolicked and played with a plush toy dog. Unlike many dogs from puppy mills, who can be traumatized and unapproachable, this little lady seemed hungry for affection, bounding into Michelle's lap and trying to lick her face.

In many cases, puppy mills are not illegal. Often, hundreds of dogs can be kept in cages for their whole life legally as long as they are provided with food, water, and shelter.

"This little dog started running around and just being very silly, and playful, and very charming, and we decided"—here Michelle broke into a huge, irrepresible grin—"that we might just want to bring her back." The little pudding-colored dog became known as Mei Mei, which is Chinese for "little sister." The youngest of the rescued dogs, Mei Mei had a name that suited her sweet, high-energy nature very well.

At the time, space at Dogtown was at a premium. The only space available for small dogs was for those that were elderly—Mei Mei was too high-energy for them, so Michelle decided to foster Mei Mei herself until a permanent adoptive home could be found. "I don't really consider myself a small-dog person, but every time I say that, I wind up falling in love with a Chihuahua!" she laughed. Mei Mei would live with Michelle at her house, provided she could get along with Michelle's cat and three dogs.

After a physical exam gave Mei Mei a clean bill of health, the Chihuahua traveled to Michelle's house for introductions. Upon arrival, Michelle

When deemed ready for adoption, Mei Mei and other dogs like her often travel to off-site adoption events like this one to find forever homes.

brought Mei Mei into the back yard to meet her dogs: Citra, a large, 50-pound black-and-tan mix; Mona, an older dog of about 40 pounds; and Espresso, a long-haired, 20-pound mixed breed. Michelle looked on as her dogs sized up the new arrival and then sniffed her all over, even following her around the yard as she investigated her new surroundings. A little hesitant and slightly intimidated at first, Mei Mei soon realized these larger dogs meant her no harm. She relaxed and began to return their greetings, smelling her new housemates with gusto.

The introductions were going well: Citra and Espresso were relaxing alongside Michelle as Mei Mei continued to explore the yard. Then they heard a sharp little bark. And then another. Mei Mei had spotted Wesley, the cat.

Ears up and alert, Mei Mei trotted over toward the large black-and-white cat, who was sitting calmly alongside the garage. When she came within five yards of Wesley, Mei Mei froze in her tracks and stared at this strange creature, who sat motionless except for the flicking of his fluffy tail. It was most likely the first time that Mei Mei had ever encountered a feline, and she wasn't quite sure what to make of him. The cat coolly stared back at the dog. Wesley was bigger than the Chihuahua, and, according to Michelle, could be aggressive with dogs if they came too close. Mei Mei smartly seemed to sense that the cat was something best left alone and turned away. "Good choice, Mei Mei! Good choice!" cheered Michelle. The introductions were deemed a success, so Mei Mei could officially move in with Michelle and begin to learn the real-life skills she would need for an adoptive home.

HOUSEBREAKING MEI MEI

Mei Mei's sparkling personality made her a strong candidate for adoption, but there was just one problem—and it was a big one. Mei Mei was not housebroken.

Like many dogs raised in puppy mills, Mei Mei had been kept in a small cage where she ate, slept, and eliminated all in the same place. Besides being unnatural and filthy, such a situation did not require that she learn to control her bodily functions. To improve her chances of adoption, Mei Mei had to learn to go to the bathroom outside. That was the primary task before the two of them when Michelle took Mei Mei home.

At first Mei Mei would go outside, busily sniff all the bushes, and then go back inside to do her business on Michelle's bathroom floor. So

Michelle started working out a routine, taking Mei Mei outside when she first got up in the morning, after she ate, and then again just before bedtime. Many dogs have a preferred bathroom pattern, and by watching her closely, Michelle realized that Mei Mei preferred to go to the bathroom after meals. So she began taking her outside into the yard right after she ate. When Mei Mei cooperated, Michelle praised her lavishly: "Good girl, Mei Mei! Nice job!" If Mei Mei had an accident, Michelle ignored it. The best strategy, Michelle felt, was to reward Mei Mei for doing what she wanted to see happen and not to punish her when she had an accident.

The routine and the praise seemed to be working for the confident little dog. Michelle kept up a consistent routine for Mei Mei, taking her outside after mealtime and encouraging her when she took her bathroom breaks in the yard. Smart little Mei Mei was catching on to housebreaking and had made enough progress to be put up for adoption. As much as she enjoyed having Mei Mei living with her, Michelle thought the little dog was ready for a new forever home. "She may need a little work on house-training when she first arrives, and she may be nervous with new people, but I think she'll adjust pretty quickly."

THE MOST POPULAR POOCH

Michelle learned that after a lifetime of little human attention, Mei Mei quickly figured out that she wanted it all the time. Mei Mei started to come to work each day with Michelle and soon had scores more fans at Dogtown because she was so charming. The dog wanted to be held constantly, and everyone was powerless to resist. Mei Mei got quite spoiled by all the attention, but Michelle still remained her favorite person. If Michelle needed to go to a meeting, Mei Mei would curl up on Michelle's chair or sometimes right on her desk, looking like a little paperweight while she waited for Michelle to return.

All of this socialization plus the housebreaking efforts had greatly improved Mei Mei's chances at finding a home. Michelle was confident that she would catch someone's eye at an upcoming adoption event held about 80 miles away at a pet-supply store. Mei Mei and a small group of other dogs were crated up and then loaded into an air-conditioned van before making the trip.

Once they arrived, Mei Mei's exhuberent personality came shining through, and she became as popular with the crowd as she had with the Dogtown staff. First, children mobbed her, then a burly man with a barbed-wire tattoo gently picked up the Chihuahua and petted her. Last of all was a sweet-looking woman named Beverly, who calmly stroked Mei Mei as she held her. Mei Mei sat contentedly in Beverly's arms, as if she knew this was a good match for her. Beverly had a 12-year-old shih tzu named Lacey at home and was looking for a perfect younger companion to complete her household. Mei Mei, it turned out, was it. "She just caught my eye," Beverly said. "She's so cute!"

An estimated 2,698,176 puppies that have originated from puppy mills are sold annually.

When Beverly got Mei Mei back to her house and introduced her to Lacey, the two dogs hit it off right away. About the same size as Mei Mei, long-haired Lacey enthusiastically accepted the younger dog and scampered after her as Mei Mei explored her new home; it was the beginning of a beautiful friendship. "When I heard Mei Mei had been in a puppy mill, it just made me furious to think that people could treat animals that way," Beverly said. "I just wanted to rescue her."

It's sobering to think of what might have happened to Mei Mei and Parker if they had not been rescued. Mei Mei, with her expressive, almost human eyes and her insatiable hunger for human attention, would probably have been confined in a tiny cell for her natural life. Parker would probably have continued to live in his ramshackle rabbit hutch with the three other dogs, though he probably would not have lived very long because the raging infection in his mouth would most likely have gone untreated.

Fortunately that did not happen. Instead, a new ending has been written for them. Parker and Mei Mei are bound for a better life than most puppies in the windows—thanks to Dogtown, they are happy, healthy, and in good, loving homes.

Exhuberant Annie had to overcome a bite history to find her forever home.

Annie: One Bad Decision

In human society, first-degree murder is the unforgivable sin—the sin above all others. A person convicted of that unspeakable transgression is often sentenced to life behind bars, or even death.

But for dogs, the bar is set much lower. If a dog bites a human, that is often considered unforgivable and punishable by death. If, even worse, a dog attacks and bites a child, retribution from human society is usually swift and terrible.

That's what happened to Annie.

Her life changed one summer afternoon when someone left the gate to her yard open and she slipped out of her suburban backyard into the yard next door. Annie was a smallish Australian shepherd-retriever-cattle dog mix, mostly black and gray, with white feet, a white-tipped tail, a white brushstroke up her nose, and a white belly. Like most "Aussies," she was high-spirited, energetic, playful, and eager to please. But when Annie was placed in unfamiliar situations, she got scared, which led her to make bad decisions.

When she entered the neighbor's yard that afternoon, she was startled to see a two-year-old toddler at play. Annie wasn't used to small children, and the toddler frightened her. Anxious and confused, she started barking at the child. When the toddler's father heard all the commotion, he burst out of the house, with an infant in his arms. He bounded down the stairs, yelling and gesturing at Annie, convinced she was about to attack the child. But the man's behavior scared Annie even more. That's when she made a bad, bad decision—one that put her life in danger.

She jumped up like a circus dog and bit the infant in its father's arms.

At that instant, Annie's world changed forever. Annie's owners, informed about what had happened, took her back to the shelter from which she'd been adopted, to have her euthanized. This incident was the first time Annie had ever bitten anyone, but it didn't matter. A first-time offense could certainly be her last.

Thousands of dogs are "put to sleep" every year for biting children and for any of a dozen other reasons. In fact, according to the American Humane Society, about five million animals are euthanized each year. Overall, about 56 percent of the dogs that enter shelters are euthanized, usually by lethal injection. But a dog entering a shelter with a bite history like Annie had only the slimmest chance of ever getting out alive.

At the shelter, Annie was essentially put on death row for dogs, waiting for an injection of pentobarbital or some other sleep-inducing barbiturate, which would cause unconsciousness and respiratory and then cardiac arrest within about 30 seconds. Annie needed a lucky break.

Rather than giving the go-ahead for lethal injection, someone at the shelter called Sherry Woodard, a Behavioral Consultant at Dogtown, wanting to know if Sherry would be willing to take Annie and try to rehabilitate her, so that she might be adopted. The shelter staff knew that Annie wasn't a vicious dog; she had been put in an unfamiliar situation and reacted badly to it. They knew that Annie would need someone like Sherry with time and resources to help her work through her fears and help her make better decisions in the future.

A HIGH-ENERGY DOG

Annie certainly didn't look like a vicious dog. She was less than knee high, with warm amber-brown eyes filled with a look of expectancy, as if she were just waiting for the next game. (Aussies are sometimes called ghost-eye dogs because a number have an eerie gray-blue eye color—but they can also have an amazing variety of other eye colors, like a kid's sack of marbles, including green, hazel, glassy blue, or, like Annie, amber-brown.)

Aussies are herding dogs developed on the ranches of the American West (not, as their name implies, Australia). Sheepherders from the Basque region of Spain brought an ancestral breed to Australia, then to America, where they developed into the modern breed. They're famous for their intelligence, trainability, energy, and eagerness to please. They are "perpetual puppies" that love to play. They excel at dog agility games, flyball (a dog sport that began in the late 1960s), and Frisbee, leaping through the air as if they were about to grow wings and fly. And normally, they're great with children, loyal, playful, and gentle.

Approximately 60 percent of dog bite victims reported each year are children, most of them being boys between the ages of six and nine years old.

Like all herding breeds, Aussies are highly energetic and need a job to do, or they quickly go a little crazy. It's as if the great open spaces of the West are built into the breed; some simulation of all that room to roam is required for them to be happy. In fact, according to the American Kennel Club, one of the most common reasons Aussies wind up in shelters is that "their owners didn't realize how much energy the breed has, and weren't willing to channel that energy through training."

But this collection of virtues, together with a fateful circumstance, may have at least partly accounted for Annie's biting incident. Because of her strong herding instincts, she tended to exhibit guarding behavior and also displayed a tendency to chase or nip at strangers. And she was flying through the air when she nipped the child, which made the bite even worse—tearing the skin in addition to puncturing it.

Sherry was informed that the infant was taken to the hospital and apparently fully recovered, though she lacked any further details. But she assumed that because Annie, the Aussie, was so "light on her feet," the bite would probably be classified as a puncture with tearing, which would be a very serious development.

(A widely used "dog bite assessment tool" developed by Dr. Ian Dunbar, a veterinarian and author, ranks the relative severity of dog bites on a scale of one to six. In level one, the dog lunges and snarls, but there is no contact with the person. In level two, the dog's teeth touch the skin but there is no puncture. In level three, the dog's teeth puncture the skin but there is no tearing or slashing. Level four is characterized by one to four punctures, with tearing, and level five by a "concerted, repeated attack." In level six, an attack results in fatal injuries.)

By Dr. Dunbar's scale, Annie's bite would rank as a three or a four.

"A serious bite to me is a puncture with tearing," Sherry said, and though such bites can be accidental, they do require medical attention and sometimes leave scarring. Sherry did not try to minimize the seriousness of the bite. Still, when she heard about the circumstances of the incident, she felt that Annie did not intend to harm the infant—she was scared, she felt threatened, she lashed out. Annie's main problem was that she lacked social skills. She was not a mean, bad dog intent on attacking and hurting people. It's possible that her owners never socialized her around small children, which led to her fear during the confrontation with the toddler and the parent. "I think she was just having a moment of panic and making bad decisions," Sherry said.

DEATH ROW TO DOGTOWN

Frankly, Sherry said, when the shelter first contacted her about taking Annie, she was terribly busy and really didn't want to get involved. "They told me the entire situation and they said, 'Can you take this dog?' And I said, 'Let me think about it.' " Sherry was hesitant—she knew it would be a big job—but after she hung up the phone, she couldn't stop thinking about Annie.

When the shelter called a second time, they turned up the heat a little bit. Sherry's friend at the shelter began hinting broadly at the difficult decision that would face them if Sherry couldn't accept Annie. It was the position most shelters found themselves in—overwhelmed, understaffed, and underfunded, they had no way of taking the time to retrain a dog, especially one with a bite history, to make her safe around people. Euthanasia (from the Greek for "good death") was the quickest, simplest, and often the most humane alternative for most animal shelters. "It wasn't like I was jumping at the chance to have Annie in my life," Sherry explained, but she couldn't help wanting to help the dog. Despite her intense work schedule and her misgivings, Sherry was won over and decided to work with Annie. After all, she said, "I'm living what I'm passionate about, and it's very important to me to feel that I'm making a difference."

That was the luckiest day of Annie's life—the day she went from death row to Dogtown.

Before Annie arrived at Dogtown, Sherry began developing a plan for retraining her. The ultimate goal was to help Annie become comfortable in a wide range of situations involving different settings and people of all ages to help her work through her fear and aggression. She wanted Annie to be as comfortable as possible so she could go out and experience life and do the things that dogs generally do. Sherry did not feel that Annie was a vicious, biting dog. She was just one that was frightened and socially inept.

Dogs that lack positive experiences and exposure to other dogs and people might also lack basic social skills that are necessary to have a healthy, happy life.

"I think that's something that people don't think about," said Sherry. "They really expect dogs to be comfortable even if they haven't been exposed to things. And out of that, dogs like Annie can get into serious trouble because of her fear, just a lack of social skills." She was, in some ways, like a nerd who hits.

If Sherry failed at this task, there was a likelihood Annie would never get a chance at a happy home. To intensify her efforts to help the little brown-eyed Aussie, Sherry not only began teaching her at Dogtown, she decided to take her home as a foster dog.

AN AMAZING GIRL

When Annie first came to Dogtown to be fostered by Sherry, it was clear from her body language that she was a very anxious dog. She kept her

Annie, an Australian shepherd, is a high-energy dog who thrives when she's able to get lots of exercise and time for play.

shaggy body low and tense, ready to react to the first hint of a threat. Annie's gaze was nervous, constantly scanning and searching her surroundings. But what was clear to Sherry was that Annie could be a very gentle dog and a great playmate if she could overcome her fear of new situations. If her energy and enthusiasm could be brought out, Annie would be a wonderful companion. Sherry saw the potential in Annie and was determined to help her shine.

Although Sherry feared that she couldn't satisfy an Aussie's constant need for exuberant play, she was pleasantly surprised to see how easygoing Annie could be. Annie settled right in to Sherry's life. She played when Sherry played and napped when she napped. As Annie grew more comfortable with Sherry, her playful sense of humor began to emerge. Sherry was pleased to see that Annie seemed "to enjoy life as much as any dog I've ever met." Annie entertained Sherry by being silly: "After she's been swimming in the creek, she'll drag her body across the ground and play in the sand and just turn herself into a sandbug, which I find charming."

Annie also began to show an independent streak and was happy to entertain herself if Sherry was busy. Sometimes, Sherry would find her playing ball by herself or romping with the other dogs, usually engaging in a game of keep-away. Annie would pick up a toy, show it to the other dogs, and then run to get the others to chase her. She clearly enjoyed her new dog friends and life at Sherry's house.

It was turning out to be a great match, and Sherry's heart began to melt for Annie. "I'm very glad Annie is in my life—she's an amazing girl," she said. "We've had a lot of fun together and we'll undoubtedly know each other now hopefully forever, for the rest of her life." The pair were getting along swimmingly, which was a great start to the hard work they had to do to overcome Annie's bite history.

LEARNING NOT TO BITE

"One of the major tools of a dog's life is their mouth—it's kind of their opposable thumb, I guess you could say," said John Garcia, Assistant Dog Care Manager at Dogtown. "And for a dog to use their mouth in a negative manner like that shows us she's never learned 'bite inhibition.'

"Or she's never learned that she can have a different warning system. Every dog has a warning system. The first level is avoidance. The next level is growling—vocalizing her emotions. The next level after that, she could show her teeth. Or she could 'muzzle punch,' which means bumping something with her mouth closed. And then air snap. And then bite."

But there's one big reason why dogs with a bite history are so often put down at shelters, John added: Doing the work required to make a dog comfortable in all situations, to make "the right choices," is a time-consuming task. In fact, all behavior modification takes time—sometimes, enormous amounts of time. And time is in short supply at shelters, as it is everywhere else in modern life.

At Dogtown, it's different, John said: "We do have that time here. Plus we have the resources, we have the knowledge, and we have the people. It just takes a lot of patience, and somebody who knows what they're doing." John was confident that Sherry could teach a dog like Annie, who had bitten once, not to bite again. Sherry herself had trained John and taught him how to succeed at getting biting dogs to stop biting, so he knew firsthand that Sherry could show Annie that she had communication options other than biting.

Quite often, John added, dogs bite because they are simply trying to communicate that they are in pain. He guessed that seven out of ten times dogs don't bite for some behavioral reason but because they're in physical discomfort. In one case, a dog came in that had bitten several people in a rescue operation, and a simple vet exam determined that he was in severe pain. He was put on medications and stopped biting. Sometimes it's that easy.

Dogs who are stimulated both mentally and physically are in general better behaved than dogs who are not. Tired dogs are less likely to get themselves into trouble out of boredom.

Other times, dogs bite people not because they mean to hurt someone but because they get overly excited and go into a "hyperarousal" state. When they get really excited like that, it's like they forget how to act, John said. They almost get "tunnel vision." It's possible that's what happened with Annie. "High arousal's probably one of the biggest problems we have at Dogtown, because it's a very stimulating environment, and when you have an environment like that, dogs do kind of get out of their heads. Then they start mouthing people. Then they start mouthing harder, which turns into a bite."

What started out as a party, winds up with an arrest.

FACING DOWN FEAR

As one of the first steps in introducing Annie to the big, scary world and teaching her to stop reacting to it by biting, Sherry took her to see a dog groomer at Dogtown. Sherry knew the groomer well and thought her easy, laid-back manner would be a good start for Annie, since even such a comparatively low-key experience could send the dog's anxiety level through the roof.

The groomer picked up Annie and set her in a utility sink, where she began washing the dog off with a hose. Ears up and alert, Annie seemed unfazed by the soap and water. The loud noise and hot air of the power dryer would be the next challenge. Annie startled at the sound of the dryer being turned on. As the hot air hit her fur, she tucked her tail and tried to get away from it, moving from one end of the deep sink to the other. Sherry was pleased to notice that Annie, though clearly frightened, was not panicking and, better yet, not growling, snapping, or biting. Sherry stepped in and stroked Annie gently as the blow-drying continued. Annie settled down but still nervously watched the dryer's every move until it was turned off.

In the next step, the groomer tried to trim Annie's nails. As the groomer held one of her paws and clipped the nails, Annie let out a high-pitched, anxious whine. The groomer had cut one nail too short, which hurt the dog. "Sorry, honey," the groomer said.

But Annie's anxiety level was rising, and Sherry again stepped in to try to calm her. Only this time, it didn't work as well. Annie bared her teeth at the groomer as she tried to work on another paw. Annie was not happy, and neither was Sherry. They decided to end the grooming session. Sherry muzzled up Annie, knowing they had a lot more work ahead of them. But Annie's choice to bare her teeth rather than to snap or bite signaled progress. Annie was learning to tell people when she felt threatened rather than attacking right away. It was a step in the right direction.

MORE TESTS

Soon after, Sherry took Annie out for a more pleasant test. She drove out into the idyllic canyon country not far from Dogtown, where a small, clear stream meandered in and out of cottonwood shade, and there were little sculpted dunes of pale sand—and no children. Then Sherry took off Annie's leash to let her roam freely and to see if Annie would come when called. Annie seemed transported into a kind of canine rapture, rambling along scent trails here and there, splashing through the creek, rolling in the warm, white sand. But when Sherry called to her, Annie immediately

Behavioral consultant Sherry Woodard observed that Annie seemed "to enjoy life as much as any dog I've ever met."

came back. Sherry was impressed that Annie had such great recall—she could follow the meandering trail of sense memories, like bread crumbs, and not get lost. She was also impressed with Annie's obedience and her quickness to return.

Sherry had been working with Annie for more than a month and was confident that she had made great progress. The next challenge would measure how far Annie had come. The two went to a public park where they could observe, from a distance, kids playing on the playground. Annie would be wearing her leash and muzzle, and Sherry would be watching her to monitor Annie's reactions. Annie had made great progress learning to control her impulse to bite, but Sherry was still a long way from feeling she could let her loose in a park.

Keeping her muzzled and controlled made Sherry certain there would be no danger to the children. "This is a situation where I have control," Sherry said, as Annie eagerly observed the herd of little children scampering around the park. "If there's ever a situation where I think she is going to hurt someone, I'm going to manage it."

It was a glorious summer day. Distantly, kids in bathing suits were horsing around in the splash park, while others did cartwheels in the grass. Looming over the little park was a red-rock butte, the sort of landscape feature so vividly evoked in Hollywood Westerns. By contrast, the grass in the park seemed absurdly, impossibly green.

The last time Annie had been close to children was the fateful day she bit the infant. Now Annie seemed extremely alert and interested, but not out of control. Holding Annie's leash with one hand, Sherry knelt at the edge of the grassy field while kids whizzed past, ever closer. Annie was standing up on all fours, intently watching, but she wasn't straining the leash. "That's good," Sherry said, soothingly.

Instead of showing fear or hyperarousal, Annie just warmly wagged her tail. She seemed particularly excited by the sight of some older kids, perhaps ten to twelve years old, wagging her tail so excitedly that her whole body wiggled. Sherry thought Annie might think those were "her" kids, the children in the family she lost.

"It's a good sign, but it's also a sad sign," Sherry said, "because she had a wonderful life I'm sure. I think she lost a lot."

Shortly afterward, a mother scooped up a small blond girl in a sundress, who was playing fairly close to Sherry and the little dog. It set Annie off, and she struggled against the leash. Sherry held firmly to the leash and helped Annie to settle down. The result of the test was clear: Smaller children still scared Annie.

"I think we will need to be very careful when parents and children are interacting near her for quite some time," Sherry said. "Maybe part of what we need to work on is that the parents are part of the picture, because Annie's bad experience included a parent. It will take time, and I want to see some different behavior before I'm confident that Annie has it all figured out."

Overall, Annie's trip to the park was a good experience. Except for the one instance of panic, Annie was relaxed and played: She lolled around in the grass, enjoying the gorgeous sunny day. She was able to watch and listen to the children at play without being scared. She became so comfortable that she was able to take her eyes off them and pay attention to other things in the park. Sherry ranked the outing to the park as a success for Annie. The pair would just need to work harder on helping Annie see that little kids were no threat to her.

THE ULTIMATE TEST

Finally, Sherry arranged a kind of "ultimate test" for Annie—a one-on-one encounter with a blond, dog-friendly two-year-old named Zoe. Sherry's friend Carragh, who is Zoe's mom, had agreed to help with the test.

"Sherry called me and asked me if I would have Zoe come and meet Annie," Carragh said. "And, yeah, I'm a little nervous about it."

"I'm very grateful to my friends for allowing this to happen," Sherry said. "Annie will be wearing her muzzle, so it will be very safe." The test began by Sherry putting a more restrictive muzzle on Annie than the one she typically wore. This new muzzle, which resembled a small basket, was made of hard plastic and fit completely over her mouth and nose so that she could not bite. Annie did not like the new muzzle at all. She pawed at it and rubbed her snout against the ground in an attempt to get it off. But her attempts were in vain. The muzzle would stay on until the test was over.

"If I've done the work that I should have, Annie should be able to get very close to Zoe and be comfortable," Sherry said. "So neither of them will be picking up stress and anxiety from the other. I think dogs are more comfortable when they're moving, so I'm going to have everybody simply keep moving. We walk and we see how it goes and if it seems like it's going really well, then we could slow down and even stop."

These isolated canyonlands near Dogtown provided an ideal place for Annie and Sherry to work off leash together.

Now Sherry, Annie, Carragh, and Zoe walked, fairly close together, out into the grassy sward of the park. Sherry had Annie on a tight leash, muzzled. But though Annie was so close to Zoe they were almost touching, Annie did not seem to be reacting at all.

"Want to pet her?" Carragh said to Zoe, gently. Zoe reached over and petted the little dog, who did not seem to mind.

"Is she soft? Say 'Hi, Annie,' " Carragh said. Zoe drew her fingers through Annie's luxuriant fur. Annie lifted her muzzle up, eyes closed, as if she wanted to enjoy it more. Then Annie shyly moved closer to Zoe and rubbed the back of her head against Zoe's stomach.

"Aw, you like her," Carragh said.

Zoe put her face against Annie's face, and Annie tried to kiss Zoe through the muzzle. She was at ease with this child and showing her affection.

"Aw, that's sweet," Carragh said. "She's doing good!"

"She's doing great!" Sherry said. "What a day!"

"It was wonderful and precious to see them together," Sherry said after it was over. "I'm really hopeful that she can get out there in society and have a lot of fun and be safe. I think if she's with the right people, the chances are she will succeed. Annie's going to be a joy to share life with."

HIGH JINKS AND A HOME

And as it turned out, Annie *did* find the right people to share her joy with, and her life. After watching the *DogTown* episode about Annie on television, a Las Vegas family decided to adopt her. A woman named Irene drove up to the sanctuary to take Annie back home, where she lived with her husband, Steve, and their 29-year-old daughter Alison.

The three of them fell for Annie's high-energy high jinks and her seemingly inexhaustible enthusiasm for play. She loved to bounce into the air like a pogo stick, and "she will jump into my husband's waiting arms, jump down and do this until my husband tires," Irene wrote in a posting to the Best Friends website (details of Annie's adoption are available at *http://www.bestfriends.org/dogtown*). "Annie never tires of this game." Annie's energy was so extraordinary—and so in need of being burned off every day—that the family took her along on their daily bike rides. Each evening, Irene would attach Annie's leash to the handlebars of her bright red bike, and then they'd all take off together, with Annie racing out ahead with so much spirit that Irene sometimes didn't even have to pedal.

Annie was especially fond of Steve and Alison, choosing to sleep on Alison's bed. She loved to go on car rides, sitting in the passenger seat just like a person, and selectively choosing to bark—or not bark—at people in other cars at stoplights. "There are definitely some people she likes better than others!" Irene reported.

An uncomfortable test took place when Irene and Steve's year-old granddaughter came to visit. While adults were present, Annie stayed in the same room with the baby for a short while, but "she kept her distance," Irene said. The little dog was not aggressive, nor was she affectionate; she merely tolerated the baby (who was never left alone with Annie). Perhaps the trauma of the incident that changed her world was still too vivid for Annie to come any closer—and that may be best for everyone.

All in all, wrote Irene, Annie "loves our family and is protective of all of us. She has some funny antics and makes us laugh all the time. We are all so glad that we adopted her." Which is a convincing testament to just how far Annie has come.

Jenny's Gift

Sherry Woodard, Animal Behavior
and Training Consultant

As anyone who comes over to my house can tell you, there is a motley crew of animals that live there with me. All my animals are special, but visitors all seem to gravitate to Miles, my eight-pound, incredibly cute Chihuahua mix. Everyone seems to want to get their hands on him and pick him up. But if Miles isn't in the mood, he may bite a well-meaning person. I find that I'm on guard for this behavior and take care to explain to first-timers that not every dog wants to petted and picked up by every person they meet every day.

Miles's zest for life and intelligence are obvious to everyone. He can show amazing bravery by running up to any other animal without a second thought for his own safety. And, despite his occasional crabbiness, he can also be astoundingly gentle: When I was fostering some baby rats in our home, Miles learned to be careful with them as if he knew they were vulnerable and needed special care. When I think back to how Miles came into my life, I think that might be where he acquired these two sides of his personality. He had a rough start, one that required both gentle care and bravery.

It all began with a call for help. A local woman was experiencing maximum stress and called up Dogtown. In addition to caring for her elderly, sick mother, the woman was left to care for two dogs and the four-year-old child of her daughter, who had disappeared. To add to the woman's woes, one of the dogs, a purebred Chihuahua, had delivered a litter of five puppies three days earlier. The mother dog had stopped nursing the pups, and the woman didn't know what to do about it. "They're all going to die," wept the woman, who was clearly overwhelmed. The receptionist at Dogtown forwarded the call to me.

I gathered supplies and headed out to her home. I thought my job that night would be simple: provide her the skills and supplies to supplement the litter with formula since the mother dog had stopped feeding them. Was I ever wrong.

Upon arrival, I walked into pandemonium. Two elderly women were wandering around the house, the dogs were barking, the four-year-old child was screaming . . . and the caregiver, the woman who had called, was frazzled and exhausted. She was definitely trying to keep it together but losing the battle right in front of me.

We sat down, and I started asking questions about the mother dog's medical history. Unfortunately, she had no history. She had never been to the vet, never been spayed, and had given birth to four litters of puppies. They had let her wander and breed, so they had no idea who fathered any of the litters. All of her previous puppies had been sold for profit. The family had a sense of pride that the mother dog had never required any additional medical attention. But many problems and sicknesses can be invisible to the average person. Mama dog could have a host of health problems that her owners would never detect. To my great surprise, they didn't see this situation as problematic at all.

At this point, I could not believe what I was hearing, so I offered to take the mother Chihuahua and her five puppies back to Dogtown. I wanted to get them to our vet to be examined as soon as possible. When I met the mother dog, I realized she had raging mastitis (inflammation of the breast) and was in pain, which I realized was why she had ceased feeding the puppies. I reached to pick her up, and, like a lot of dogs who are in pain, she bit me.

I asked the woman to find a box and bring it to me. Then I carefully loaded the mom and her five mouse-size puppies into it. We agreed that I would find homes for any puppies who survived, and I would return the mom after she was spayed.

I was (and still am) lucky enough to work at Best Friends Animal Society, where the vet staff, after a quick phone call, was waiting for my arrival. Driving to the clinic, I realized that I would not be going home that night. I was the manager of Dogtown, and there were many nights when I did not go home because I needed to care for a newly admitted dog around the clock. In fact, I considered my office a second home.

Because my office already contained animals, I asked Faith Maloney, one of the founders of Best Friends, if I could borrow her office for the mother and her pups. As it turned out, she lent it to me for the next few weeks. I moved in too to care for the pups!

When we arrived at the clinic, the veterinarian found that the mom had extreme mastitis and was running a high temperature. Her parasite load was so heavy that she had shared it with her pups, and they were passing blood instead of stool. Two out of the five puppies were barely hanging on to life. We fed all five with syringes, since they were too weak to suck from bottles. I provided warm compresses in between feeding the puppies every three hours around the clock. At this point, the mama dog was still not my friend—she was still snapping at me.

Mom was put on medication. Within a few days, she began to feel better and three of the puppies were back to nursing part of the time, supplemented with bottle-feeding. Two puppies remained very weak and were being syringe-fed by me around the clock. Mom, now answering to the name

Jenny, had begun to trust me and made it extremely clear that she would no longer care for the two ailing puppies. She started removing them from their bed and placing them on the cold office floor. Whenever I found them there, I picked them up and gently placed them back, beside their mother and siblings.

Despite all my attempts to keep the two sick pups in the bed with their family, Jenny persisted. I starting having conversations with Jenny about her puppies, but I would still return to find the two sickly pups on the office floor. So I set up a second bed with a heating pad and a beating-heart toy, which one of our members had donated, for the two smallest pups. By this time, each of the puppies had a name. The largest boy was named Bernard, after one of our trustees in Dogtown, and the smallest was named Traveler. I named the two girls Cascade and Pixie. The remaining boy, Miles, was named after one of my personal favorite dogs in Dogtown. A kind a gentle shepherd mix, Miles the elder was the epitome of a good dog but often overlooked by adopters.

Despite my best efforts and numerous visits to the clinic, the two weakest puppies, Traveler and Miles, were struggling. At one point, I fed them and left the office; when I returned about a half hour later, I discovered that Traveler had lost the fight and was gone. Miles was still hanging on, and I had to find a way to save him.

Miles, who had an open fontanel (open space between the bones of his skull), had been deemed different by his mother. Even though he looked so small and alone, I knew he could make it with his mother's help. It was at that moment I decided to start negotiating with Jenny again. This time, as we talked, I promised her that if she would allow Miles time with her and with his siblings, I would care for him for life. My true belief was that she knew he was special, and if she understood that he would be cared for, she would accept him.

That day, I was a good dealmaker. Jenny never put Miles out on the floor again. As the weeks passed, I was amazed at the change in physical appearance of each puppy. They grew healthier and stronger, but as there appearances began to change, there was no doubt that the mom was allowed to run around: Mom was a purebred black-and-tan Chihuahua. Both surviving males, Bernard and Miles, were becoming long-haired. Pixie and Cascade were wire-haired. Bernard was the biggest, and Miles grew larger than Pixie, who was now the smallest and was taking on the appearance of a werewolf.

Even though I knew the puppies wouldn't be ready for spay/neuter for quite some time, I started screening homes for them. At Best Friends, we do spay/neuter when dogs are at least eight weeks old and weigh over two and a half pounds. It would be a while before these puppies reached that weight.

I contacted Jenny's original caregiver to give her an update on the family's progress. Her own personal situation had not improved enough to take Jenny back into her home, so I offered to find Jenny a new home as well as homes for all the surviving puppies. She agreed that would be the best situation for Jenny and her litter.

It didn't take too long for Jenny and her family to all find wonderful places to live. Jenny and her daughter Cascade both went to live with a Wisconsin family who had adopted from Best Friends in the past. Pixie was adopted by a staff member, and Bernard, still growing fast, was adopted by a Best Friends founder. As for Miles, I honored my promise, and it couldn't have worked out any better for me. It turned out that he was very special. His cuteness, his spirit, and his bouncy manner have made him a fun addition to the animals who live at my house.

To this day, I am thrilled to give him the constant supervision and protection that Miles requires. For the most part, he has enjoyed good health since his difficult delivery into the world. His size makes him easy to manage—I can scoop him up if other animals are approaching. I love Miles very much and think of him as a gift from Jenny. I hope to enjoy sharing my life with him for many years to come.

Because Best Friends Animal Society provides a lifetime commitment to every dog adopted from Best Friends, including Jenny and her pups, we are able to keep up with adoptive families and arrange the occasional family reunion. One Mother's Day, Jenny and Cascade traveled back from

Wisconsin to reunite with their family. It was a happy reunion, and I had the chance to thank Jenny for striking a deal with me all those years ago. I wanted her to see that I had held up my end of the bargain—that Miles was healthy, safe, and loved.

Tuffy's spirit never flagged after Dogtown rescued him from a hoarder.

Tuffy: The Will to Live

If a person were trying to find the smack-dab middle of absolutely nowhere, Gabbs, Nevada, would have to be close. It's the sort of place people go in order not to be found, and generally succeed. It is a flat, relentless emptiness not far from such scenic wonders as Cactus Flat and the Humboldt Salt Marsh, though distantly one could see eerily beautiful, dead-dry mountains, like mountains on the moon. It was here that Dogtown rescued a resilient young Dalmatian mix they named Tuffy. Tuffy never gave up on life, and Dogtown never gave up on him. It was the perfect combination to bring the young dog back from the brink.

THE PROBLEM OF HOARDING

In December 2007, Dogtown got a call about a horrific hoarding situation near Gabbs. A woman contacted Dogtown after her aunt, who had been keeping hundreds of dogs on her isolated property, passed away. The deceased woman had started out as a rescuer, taking in abandoned dogs. But the situation overwhelmed her, and she wasn't able to look after all the animals properly. Conditions at the property deteriorated, and the dogs were surviving in terrible circumstances.

Animal hoarding is a complex issue, and sometimes hard to recognize from the outside. Often people who hoard animals believe that they are helping and protecting them, which leads them to take in too many. As the number of pets grows, the hoarder becomes overwhelmed and unable

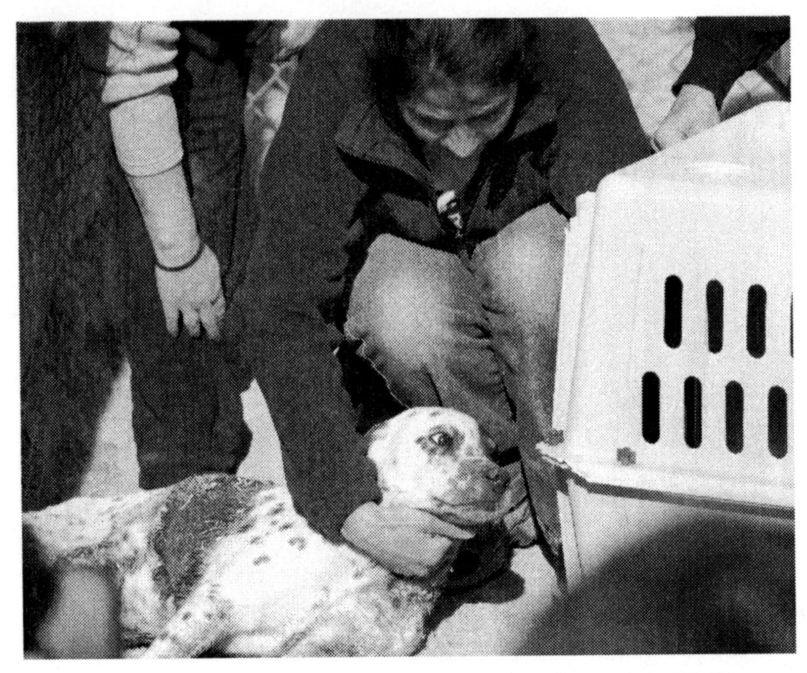

Dog Care Manager Michelle Besmehn eases the wounded Tuffy into a crate to transport him to safety.

to provide the minimum care necessary: food, sanitation, shelter, and medical care. Beginning with good intentions, hoarders end up abusing their animals through neglect. Such behavior is remarkably common: Researchers at Tufts University have estimated that there are 700 to 2,000 new cases of animal hoarding in the United States every year.

In one typical hoarding case investigated by the Humane Society of the United States (HSUS), a couple in Maryland started a shelter for cats called Chubbers Animal Rescue, complete with cheerful website. Though the couple's original intentions may have been good, when HSUS volunteers entered the house in May 2003, they found a total of more than 300 cats, of which more than 70 were dead. In one part of the house, volunteers found themselves stepping on heaps of feces and skeletons. "It was disgusting," said one volunteer. "The amount of filth was unbelievable."

Yet, like many hoarders, the couple was convinced they had been caring for the animals properly, and seemed in complete denial that their animal "refuge" had turned into a house of horrors. The pair were well educated and well spoken, and had the uncanny ability to attract sympathy to their point of view. Their justification was that they loved animals intensely, and they were afraid the cats would be euthanized if they went to a shelter.

Psychiatric studies at several institutions have suggested there is some greater mental illness involved in hoarding. The most likely candidate is a form of obsessive-compulsive disorder in which people develop an overwhelming sense of responsibility for something like taking care of animals even after they are no longer capable of doing so.

A SURREAL SITUATION

A small rescue team from Dogtown flew down from Utah into a tiny airport in the Nevada desert, drove almost two hours, then took a dusty dirt road 13 miles into the desiccated backcountry. They would be working on-site with other rescue groups to help assess the dogs and network them to other groups where they could be adopted. When the teams finally arrived at their destination, the scene that greeted them was "surreal," Dog Care Manager Michelle Besmehn recalled.

In this weird, alien landscape, with its pitiless desert sun and sense of utter desolation, was a decrepit house and a rundown collection of kennels and cages containing 150 dogs. The dogs—multiple breeds, ages, and states of neglect—all seemed to be howling at once. Most of them were in individual kennels, while several others were running around in small packs in larger enclosures. Outside the cages, a few dogs simply roamed freely. There were ratty-looking dogs trying to jump the fences, and dogs attempting to mate with other dogs. It was chaos.

When Michelle entered one of the larger enclosures where a pack of dogs had formed, she found a scene that stabbed at her heart. Among the cacophony of the other dogs, a young Dalmatian mix, not much more than six or eight months old, lay quietly on the dirt, immobile and still.

His dirty fur bore the signs of a savage attack by the other dogs; his side and back legs were covered with gaping, open wounds and slathered in dried blood, mud, and excrement. Too weak even to raise his head, his shallow panting was his only detectable motion.

Michelle examined the dog and noticed that his heartbeat was weak and rapid. His gums were pale. His skin was cool to the touch. And then there was his smell: the putrid odor of infection. Michelle knew those signs meant the dark rider of death was on its way, and due to arrive in a matter of hours. Whether the little dog actually knew this, it's impossible to say. He looked pleadingly up into Michelle's eyes, and she simply could not resist this canine soul, so young and so full of suffering.

She sat down on the dusty ground in her blue jeans, cradling the little Dalmatian's head between her legs and comforting him. Despite his injuries, it was easy to see that this little guy was a cutie. He had a sweet and comical coloration, with black ears, a scattering of white spots on the right side of his face, and a big black patch over his left eye. Dalmatians are sometimes called plum pudding dogs because of the random way blobs of black are scattered over their skin, and the big black blob over one eye had a particularly striking appearance. Michelle could see that the dog was in shock, but she also felt that he seemed grateful for the attention and the thought that help might have arrived at last.

"For me it was important for him to know that someone cared whether he made it or not," Michelle said later. When she said this, remembering the moment, her voice broke and she brushed away tears.

"I think that what struck me about him was the way he *looked* at me. You know, here was a little dog who had been injured, basically left for dead, and his eyes were actually pretty clear and he still wanted to *try*. His little tail wagged, and—I don't know—I felt like he had a lot of drive to live."

A RACE AGAINST TIME

The Dalmatian pup's only chance of survival was immediate medical attention. But there was no veterinarian with the Dogtown team, and the nearest vet was almost two hours away. Given Gabbs's extreme remoteness, there was also no cell phone service, so there was no way to call ahead to let the vet know a dying dog was on his way.

The only solution was to load the dog into a vehicle and hightail it across the desert in hopes that the vet would be available when they arrived. That job fell to vet tech Jeff Popowich, an immense man with a permanent three-day growth of stubble and an incongruously soft voice, who loaded the injured dog into a carrying crate and hit the road in a race against time.

"I was scared this was a dog that was not going to make it," Jeff said. "Just the smell and the sight of those wounds—it was bad. I was really worried about infection, that was the big thing. He just wasn't responding well, he had a fever, he was swollen all over. It did not look good."

But when Jeff arrived at the vet's office, the animal doctor, unfortunately, was gone. A female assistant let Jeff into the office, and he lifted the carrying crate with the dying dog inside onto an examining table. "Whew—that's some strong smell there," Jeff said to the assistant, almost apologetically. Jeff sighed, seemingly overwhelmed. "On the inside of his groin he's got two more big holes. They're really foul."

Animal hoarders are people who, besides having more than the typical number of pets, are unable to provide the most basic animal care, which includes sanitation, nutrition, shelter, and veterinary care.

But because there was no veterinarian on site, Jeff and the assistant were almost helpless—they couldn't administer pain medications, couldn't start an IV line for fluids, or do much else.

"There's nothing we can do right now, nothing except stay here and stare at him," Jeff said, impotently, peering into the crate where the young Dalmatian lay sprawled. Death was on its way, and there was little Jeff could do to stop it.

Rather than sit and wait for the vet, Jeff decided to help the dog in whatever way he could. He got some paper towels and medical disinfectant and began removing the caked-on mud and excrement from the dog's fur and wounds. It seemed a weak gesture, but there was another kind of medicine he was also administering: a healing touch. The pup with the black patch over one eye slowly lifted his head and looked up at Jeff, licking his chops, and then weakly lay back down. The dog's small gesture seemed to indicate that he appreciated what Jeff was doing for him, even though there was little the dog could do to show it outwardly. "Hang in there, buddy," Jeff said. "Maybe this'll make you feel a little better."

While Jeff cleaned up the dog's fur, an idea for a name came to him: This Dalmatian pup was a fighter, a tough animal with an invincible will to live. He had been through terrific pain and injury in the dogfight, and there was no telling how many hours or days he'd been lying there before he was found. Yet he was still—barely—alive.

Later that night, Michelle called Jeff to find out about the Dalmatian pup. Had he made it? When Jeff answered his cell phone, he told her that the vet had finally arrived, cleaned the little dog's wounds, administered pain medication, and hydrated him with intravenous fluids. So far, at least, the puppy was alive, and they hoped he would make it through the night. Oh, and by the way, Jeff said: The tough little puppy's new name was Tuffy.

DR. PATTI TAKES OVER

Tuffy made it through the night and then made it all the way back to Utah. When Tuffy was brought to Dogtown, it was with some other dogs rescued from the hoarder in Gabbs—a group of puppies who seemed to be healthy and bouncing around, and a dog named Toey that had a strange mass on his groin. But Tuffy was the one that needed the most urgent attention, and the one that Dr. Patti Iampietro, one of the Dogtown vets, looked at first.

When Tuffy arrived, he'd been cleaned and bandaged and he had an IV catheter in place.

Even so, Tuffy was "very, very subdued, especially for a puppy," Dr. Patti said. He just lay quietly on the table, his limbs swollen, his body listless. Still, at least he was somewhat responsive—he wasn't comatose or in a state of severe shock. Dr. Patti credited this to the work of the Nevada vet.

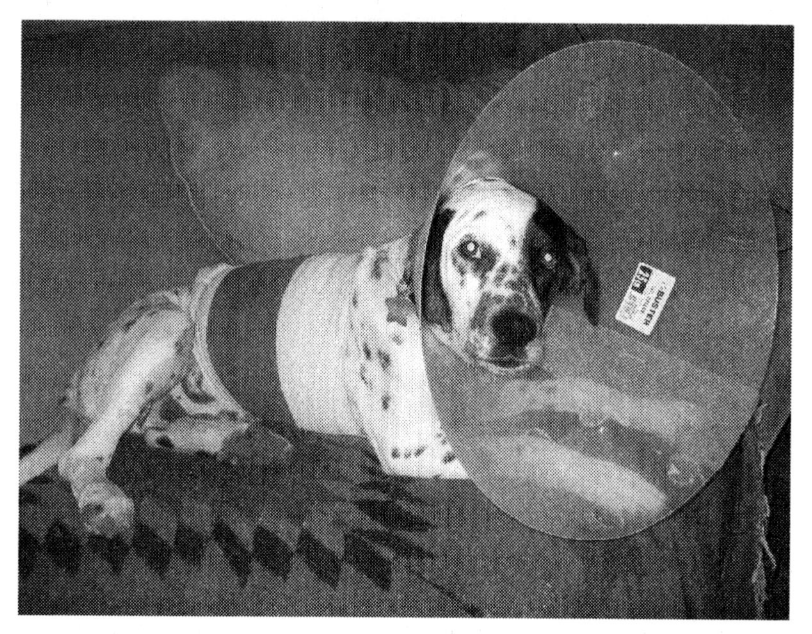

An oversized plastic collar around his neck and snug bandages around his middle both protect Tuffy's skin as it heals.

But Tuffy was clearly in distress. Ten to fifteen percent of his body was an open, infected wound that was several days old. So Dr. Patti's first mission of mercy was to give him pain medication and make him comfortable. Then she slowly, carefully peeled back the bandages. Underneath them she found what she later described as "horrendous wounds."

12 to 15 inches long, which had been pulled open all the way down to the underlying muscle. The wound was gray and smelled foul and necrotic (full of dead tissue), and the edges of the skin were starting to blacken and die. The other two wounds were three- to four-inch-long gouges on the insides of his back legs, very deep, also gray and necrotic, with black edges. There were three major lesions. The largest was an enormous tear on Tuffy's right side,

Even after the Nevada vet had cleaned them somewhat, the wounds were grossly contaminated with dirt, debris, and dead tissue—like the wartime wounds of a soldier who had not seen a medic in much too long. Normally, Dr. Patti explained, these tissues should be "nice and pink," and in a fresh wound they would be bleeding. But their brownish gray color accompanied by the putrid smell meant the tissues were dead or dying. "Gray is definitely not a color you want to see," she said.

After graduating from vet school at Pennsylvania State University, Dr. Patti had worked for ten years as an emergency room vet, and she had sometimes seen wounds this bad—after dogfighting confiscations or after animals had been hit by cars. But what she'd never seen before were severe wounds that had been neglected and left to fester this long. If the Dogtown team had not intervened, it was certain that Tuffy would have died of his wounds.

Still, she believed that Tuffy now had a better than even chance of survival, if only for two reasons. One, he was young, and young animals have a remarkable capacity to heal after injury. And two, like Jeff and Michelle, Dr. Patti had seen something in Tuffy's eyes that might well make all the difference: his ferocious will to live.

ADDRESSING THE WOUNDS

"When I'm faced with wounds such as Tuffy's, the instinct is to jump right into those wounds," Dr. Patti said. But she did not know this little dog, did not know how stable he was, or how he might respond to the trauma of having his lesions surgically cleaned. So Dr. Patti wanted to go slow. She would take it one step at a time and assess all his vital signs—temperature, heart rate, and pulse—to make sure that he was stable before trying to heal Tuffy's injuries. "Wounds like this take a *really* long time to deal with," Dr. Patti said. "I can't just clean it up and sew it up-I'll never get all that infection out at once."

After he had stabilized, Dr. Patti treated Tuffy's wounds in careful stages. After her first assessment of Tuffy's overall condition, she began by putting fresh bandages on his wounds with some sterile saline. The bandages were special debriding bandages that, when removed, would help to gently pull away some of the gray, infected tissue. Then she put him on some intravenous antibiotics and tucked him away in a crate in

Vet tech Jeff Popowich and Tuffy shared a special bond that strengthened as the dog recovered from his wounds.

the clinic with a pain patch. If he was stable the next day, she'd consider anesthetizing him and addressing his terrible wounds more aggressively.

The next day, to Dr. Patti's delight, Tuffy actually stood up and greeted her, tail wagging weakly, when she came to visit his cage. It was a great sign. And when she put him on the table and pulled back the debriding bandages, Tuffy's wounds looked a little bit cleaner and pinker. "See, it looks better than all that nasty gray stuff yesterday," she said to Tuffy.

Then she gave the little dog a physical—checked his heart rate, his temperature, his gum color, drew a little blood. It all looked good. He seemed to be rebounding nicely, and Dr. Patti felt he would be stable enough to withstand anesthesia and surgery.

"Ideally, should we wait a little bit longer until he's stronger to do surgery? Yes," Dr. Patti said. "But I'm very concerned about his infection also, and I think [the benefit of] removing the infected tissue outweighs the risk of the surgery."

According to the Humane Society of the United States, "The general consensus is that animal hoarding is a symptom of psychological and neurological malfunctioning, which might involve dementia, obsessive-compulsive disorder, post-traumatic stress disorder, and attention-deficit hyperactivity disorder."

When Tuffy was anesthetized on an operating table, his small pink tongue was clipped so he didn't swallow it. A tube was slipped down his throat, attached to a green bag that inflated and deflated with each breath. Dr. Patti peeled back the latest bandages and began working to cut away the gray, necrotic tissue from Tuffy's wounds. She kept an eye on monitors to keep track of his blood pressure, heart rate, and oxygenation levels. Dr. Patti noticed a slight arrhythmia on the electrocardiogram, but other than that, Tuffy seemed to be coming through the procedure like a trouper.

With Tuffy fully sedated, Dr. Patti was able to get a better sense of his awful wounds. She found that she could stick her entire hand under the flap of skin on his side, since there was so much "pocketing" where the skin tissue had literally been ripped off the underlying tissue. The good news was that, even though the wounds on his side and legs were deep and ugly, they had not penetrated any major cavities in Tuffy's body.

With the puppy stretched on the operating table, Dr. Patti said later, she had to fight a strong instinct to just clean him up and sew the wounds closed. But that would be a mistake. The wounds were still so dirty, and so infected, that if she closed them, the infection would continue to fester underneath the skin and could burst open later like an overcooked pie. Instead, she patiently planned several more days of bandage changes, and medical management with antibiotics. Once the wounds looked really clean, then she'd close them up. This strategy was not without a different kind of risk, though: The longer the wounds stayed open, the greater the risk Tuffy had of contracting other, more serious infections. "He's certainly not out of the woods at this point," Dr. Patti said.

INFECTIOUS HAPPINESS

When Tuffy groggily came awake after the surgery, he found Dr. Patti cradling his head in her hands. "Hi, handsome!" she said. "I told you I would be here!"

Tuffy sported a bright blue bandage around his middle, looking as if he'd just stepped out of the shower at the spa. Although he came through the surgery with flying colors, the post-op would still be extremely painful. Dr. Patti and an assistant lifted Tuffy on a blanket into his crate, where he whimpered in a high-pitched whine, confused about where he was and why it hurt so much.

Dr. Patti ordered another dose of morphine. And she had a comical plastic collar, shaped like a megaphone, fastened around Tuffy's neck to keep him from chewing on his bandages. After Dr. Patti closed the door to his crate, Tuffy lay back on the mat. Then he reached over and tugged his blue blanket up around his chin, tucking himself in like a child home sick with the flu.

After a few more days of bandage changes, Dr. Patti had grown increasingly optimistic about Tuffy's chances. He was healing "remarkably fast," she said. "What I thought was going to take weeks is taking several days." Tuffy's recovery, in fact, was turning out to be one of the quickest transformations in Dogtown history.

During his recovery, Tuffy's optimistic personality began to shine through. Covered in his bandages, he still wanted to prance around and wag his tail despite being in pain; his spirits were high even though his legs were swollen, as though he knew what a lucky break he had caught. Tuffy's joyful disposition won the hearts of everybody he met, making him a favorite at the clinic. One of the volunteers described him as "the best dog in the world. You just take one look at him and you fall in love with him." What stood out to Dr. Patti was how Tuffy "holds no grudges. Regardless of what has happened to him, he still loves people." His tough spirit had not been broken by his history. He'd been well named indeed.

Tuffy's valiant and likable spirit could well have come from his bloodline. Dalmatians have a proud, dauntless history, having originally been bred in Dalmatia (a part of Croatia) as dogs of war. They guarded the borders of their country, and to this day still have a strong guarding instinct. But their most important role was as coach or carriage dogs, running alongside horse-drawn buggies with stately and athletic grace. Later Dalmatians' instinct to run in attendance with horses transferred to horse-drawn fire engines, which is why they are still known and loved as "firehouse dogs."

One can tell the approximate age of a dog by checking the wear on his incisors. This method is somewhat accurate until the age of six, but beginning at seven years of age, it's unreliable.

As part of Tuffy's treatment, Jeff Popowich started taking Tuffy home with him almost every night, partly to make sure his bandages stayed on and were sterile, and partly because human company was healing for Tuffy—the same way canine company was healing for Jeff. "I love that little guy," Jeff said, his big face lighting up. "He's a cool dog!" Even though he was still bandaged and healing, Tuffy was "an easy pup—he loves meeting other dogs, he loves meeting people. His happiness is infectious."

ON THE MEND

After six days at Dogtown, Tuffy had reached an important milestone. He'd healed enough to have his wound sewn shut. Dr. Patti was still concerned that there might not be enough healthy skin to stitch the wound closed, after so much had been lost to injury and infection. If she couldn't close the wound, there were other options, like skin grafts, but Dr. Patti was hoping that wouldn't be necessary. She couldn't really tell what the situation would be until she got Tuffy on the operating table and took a look inside.

When she did get in there, she found to her relief that she was able to suture the wound, though she worried that a smaller, secondary wound raked open by dog teeth might later partially break open. She'd close up the wound as well as she could and just keep an eye on it. Dr. Patti also implanted several small, flexible, sterile tubes, or drains, in the wound, to allow fluid from the wound to drain off, rather than accumulate under the skin.

A few days later, after a few more bandage changes, she opened Tuffy's dressings and found that the tissue around the stitches had returned to a clean, healthy pink. "Oh, that looks beautiful!" she said. "Well, maybe not to the average person, but I'm happy with that!"

Dr. Patti had played a critical role in rescuing a dog from the precipice of death—a dog who in almost any other animal shelter would have been dead within a matter of hours or days. Most ordinary shelters, overwhelmed and underfunded, could not offer the time, attention, and heroic effort that led to Tuffy's survival—much less the free, top-notch vet care and loving post-op recovery, which amounted to the difference between life and death.

But Tuffy was not the only doomed dog who came to Dogtown and was healed. One of many other dogs that came to mind, Dr. Patti said, was a little dog named Gilbert, who was named after the police officer who brought him in after he'd been hit by a car. He was in shock, he had a fractured pelvis, a fractured femur, bruising on his lungs—he was basically dying. Dr. Patti was called in after hours, and she and an animal control officer worked together for hours to bring this little dog around, giving him fluids and pain medication.

"But slowly I saw the little personality in this dog come out. I just thought, wow—I can fix this dog! He deserves a chance. Eventually he was

Tuffy's future family found their way to adopting him after seeing the National Geographic Channel show DogTown.

adopted into the best home and they are in love with him now. This dog, in my opinion, wouldn't have made it anywhere else in the country."

Which is why Dr. Patti, and Jeff Popowich, and Michelle Besmehn, and everybody else at Dogtown do what they do. And why Tuffy was so lucky that they found him.

HEADED HOME

Tuffy found his forever home thanks, in part, to the National Geographic Channel show *DogTown.* During its first season, an Arizona family was so touched by an episode that they went to the Best Friends website to learn more the organization. There they found Tuffy's picture and his story. The mother, Jody, said she was brought to tears by Tuffy's tale—which was no small thing, considering that she was a former animal cruelty investigator for the Arizona Society for the Prevention of Cruelty to

Animals and had seen her fair share of sorrow. The family decided to file an adoption application for Tuffy, and it was accepted.

Then the family—Jody and her husband, her daughter, and two sons—drove all the way from Arizona to pick up Tuffy at Dogtown.

"Tuffy, you're so much bigger than I thought!" Jody laughed through her tears, when she first met him in the clinic lobby. Tuffy couldn't stop licking her, as if she were a big human-size lollipop. "Oh, man!" she said. "I can't believe I'm crying!"

Then, to her teenage son, she said, "It was a long drive to come get him, wasn't it?"

"It was worth it," the boy said.

Tuffy sprawled on the floor of the clinic while Jeff Popowich pointed out the still healing scars to Tuffy's new family.

"When I first saw Tuffy, I didn't even know if he was going to live through the day," Jeff told them. "But now he bounces around, playing just like a puppy should. Compared to where Tuffy was when he came in here—it's a complete 180. Now he's a different dog."

The smaller boy asked Jeff where Tuffy usually slept. "He sleeps with me in the bed every night, right next to me," Jeff said. "That's OK, all our dogs sleep in the bed too," the little boy's older brother said.

Everyone walked out to the car, and then Jeff knelt to say goodbye to the tough little spotted puppy. Tuffy licked Jeff's face, not realizing it would probably be for the last time.

"Thank you for what you did for him," Jody said to Jeff, reaching up on her toes to get her arms around the big man.

"Spoil him rotten," said Jeff.

Then the family got in the car along with Tuffy, the little Dalmatian with the invincible will, and drove off toward a new life.

Anxiety issues plagued Rush after he survived the 2006 shelling of Lebanon.

Rush: Dog of War

The original Rin Tin Tin was a shell-shocked German shepherd pup found by an American serviceman in a bombed-out dog kennel in France, a couple of months before the end of World War I. The serviceman, Lee Duncan, named the beautiful dark-eyed dog after the little puppet that French children gave the American soldiers for good luck. Duncan took the pup home to California, where a film producer saw him performing at a local dog show. The rest, as they say, is history. Rin Tin Tin starred in a series of hugely popular films (where he often played a wolf) and is said to have died in the arms of screen goddess Jean Harlow.

Almost a century later, another shell-shocked dog was rescued from a bombed-out animal shelter in Beirut during a conflict between Israel and the Shiite group Hezbollah. During what came to be called the 2006 Lebanon war, 34 days of shelling killed over a thousand people and displaced 1.5 million Lebanese and Israeli civilians. As in every other war, huge numbers of animals were also among the victims, whether they were pets in bombed-out homes, livestock on farms, or homeless animals in shelters.

War had not changed much in a hundred years, at least not for animals. But like Rin Tin Tin before him, this dog was lucky enough to find rescuers who would help him make his way from the bombed-out streets of Beirut to the peaceful Utah canyons of Dogtown.

RESCUING RUSH

Many of Beirut's war zone animals were rescued by a small group of heroic, unpaid volunteers who called themselves Beirut for the Ethical Treatment of Animals, or BETA (no relation to the U.S. animal rights group PETA). It was Lebanon's only animal welfare organization, desperately trying to save the innocent as the bombs fell.

One of BETA's two founders, Maggie Shaarawi, who had a full-time job at the United Nations in Beirut, blogged from the war zone: "Once I finish my job, I do the rescue work and go to shelters. We do everything. We work until midnight. We don't have personal lives." Why did they do what they did? The reason was simple, Maggie said: "It feels so good to save a life."

Some veterinarians suggest dogs do not suffer the psychological distress of losing a limb the same way a human does. The primary purpose of the limb is in movement. Because dogs do not need to perform fine-motor skills, most easily adapt to having only three legs.

One of the animals they saved came to be known as Rush. A beautiful shepherd mix with golden, almost sun-blond fur, a dark, pointed muzzle, eyes so dark they were nearly black, and delicate, expressive brows, this dog was constantly on the alert, scanning his surroundings for potential threats. When he was first rescued, Rush's front left leg was badly injured, and BETA vets determined it was too damaged to save. They amputated his leg, leaving him with three good ones. But Rush's lack of a leg didn't slow him down; his fear did. Rush's up-close experience with the sights and sounds of war left him terrified, startling at loud noises and sometimes biting in response to fear.

There were only a dozen people in BETA, all doing this selfless work in addition to their day jobs. They rescued not only abandoned or escaped dogs and cats but workhorses, donkeys, and zoo animals. To raise money, they would buy movie tickets in bulk and sell them in the street and hold fund-raising dinners (mainly for their own families), and they maintained an email list of about 200 supporters.

So when a well-known, well-funded, and well-staffed organization like Best Friends Animal Society volunteered to help with the animal rescue in Lebanon, the brave hearts of BETA were overjoyed.

BEST FRIENDS REACHES OVERSEAS

Michelle Besmehn, Best Friends Dog Care Manager, was part of the team who flew to Lebanon to help with the animal rescue operation in the summer of 2006. The effort was complex, expensive, and dangerous—in fact, it was the largest international rescue operation Best Friends had ever attempted at that time. The primary mission was not to rescue dogs and cats off the streets, but to help BETA by taking animals out of their shelter, because their cramped space was filled to capacity, there were no adoptive homes available, and countless more dogs and cats were still in need.

In some ways Michelle's job forced her into the uncomfortable position of having to play God, choosing which animals would be taken back to the sanctuary and which would not. (A Best Friends committee assisted in this difficult task, but it was always heartrending to leave an animal behind in a war zone.) When Michelle and the team got to Lebanon, they found the animals, mostly dogs and cats, in various conditions. Some of the animals, mostly the ones who had been at the shelter the longest, were in good condition, but more recent arrivals were not doing well at all. Some had fresh, open wounds. Others showed signs of past illnesses, like distemper, which can cause neurological problems, or old injuries that had not healed properly, and others had new illnesses. Michelle remembers the whole experience as "very rushed, very exciting and kind of scary," not only because they were in a war zone, but because there was so much work to do.

"We had to get the dogs out to vaccinate them and do blood draws and deworming and microchipping, so it was quite a production because the shelter did not have electricity or running water. They had to haul all of their water and they had to use generators for electricity. We had to just make do with what we had."

In the process of choosing which animals to bring back, Michelle could not resist the three-legged shepherd mix with the expressive eyes, the regal stance, and the tawny, sun-blond fur. He was the sort of dog that Best Friends was specially suited to help. For one thing, his leg, fairly recently amputated, could potentially pose a medical problem, and Best Friends was superbly equipped to deal with veterinary challenges of all kinds. For another, he was a beautiful dog who most likely would be highly adoptable, and placing dogs in adoptive homes was the ultimate intent of the sanctuary. Michelle couldn't help but notice that he also seemed extremely shy, perhaps shell-shocked, and she knew that Dogtown's staff of trainers could often work magic with traumatized animals like this.

Ultimately, the Best Friends team loaded up almost 300 crateloads of animals (about as many dogs as cats) for the 7,000-mile flight home to the U.S. It was very much like a military operation, large in scale and stripped of frills, except that the cargo plane wound up filled not with soldiers, arms, and explosives, but with an amazing variety of animals, peering out of their crates with a mixture of alarm and curiosity.

TURNING OFF THE MENTAL MOVIES

When Rush arrived at his new home at Dogtown, he was withdrawn, fearful, and mistrustful. Loud noises terrified him. He startled easily. And worst of all, he was still inclined to bite when spooked. Even though he was galloping around on only three legs, his most grievous injury was not visible to the eye. It was a wound to his psyche.

Rush's symptoms of fear resembled those of post-traumatic stress disorder (PTSD), a condition that many humans suffer from after living through a traumatic experience. Dr. Frank McMillan, Director of Well-being Studies at Best Friends and author of the book *Unlocking the Animal Mind,* often sees psychological problems in animals that resemble PTSD in people. He defines the condition as "a type of emotional scar, caused by such traumatic events as combat, sexual abuse, crime, and natural disasters. Humans suffering from PTSD generally fear and

Dogtown's trainers observed that encounters with unfamiliar places, new people, and loud noises all frightened Rush.

avoid whatever caused the trauma for months and sometimes years after the event, often experiencing recurrent emotional episodes that can be severely distressing. They also may have difficulty sleeping and have frequent nightmares."

One of the reasons these traumas are burned so deeply into animals' brains is that fear serves an evolutionary purpose, Dr. Frank explained. "When the ancestors of our modern dogs and cats encountered such terrifying situations as an attack by a predator or a natural disaster, only those who acted quickly to save their own lives survived, while the slower ones perished. This resulted in greater survival of those animals with brains best able to form enduring memories of danger. For these animals, when something similar happened in the future, their fear memories would immediately motivate the animals to take action."

In other words, if Rush were still living in the war zone, his fearful reaction might save him from a bomb blast. But Rush was no longer under fire; his mind had just not caught up to his circumstances yet. Rush's survival reaction wouldn't turn off, and he continued to live in the world of war that he had left behind.

As a way of explaining why these traumatic experiences can be even more crippling to animals than to humans, Temple Grandin, the autistic author of a series of remarkable books about animal behavior, writes that "mental images are far more closely connected to fear and panic than words." And since animals do not remember traumatic experiences as a verbal narrative, as some (but not all) people do, but instead as a kind of vivid mental movie, they are easily traumatized by scary experiences—sometimes for life.

In her book *Animals in Translation,* Grandin describes several scientific studies that lend support to this claim. In one study by Ruth Lanius, an assistant professor of psychiatry at the University of Western Ontario, Lanius scanned the brains of 11 people with PTSD from traumatic experiences. Then she scanned the brains of 13 people who had suffered the same experiences but had not developed PTSD. The biggest difference was that, when people with PTSD remembered the trauma, areas of their brain associated with visual experiences lit up. When people without PTSD recalled their trauma, verbal areas lit up.

It may be profoundly true, in other words, that "a picture is worth a thousand words," even when it comes to animals. The bottom line is that "a visual memory of a scary thing is more frightening than a verbal memory," Grandin observes, even though "no one knows why or how words are less frightening, or how this works in the brain."

Whatever Rush had witnessed in Beirut may have formed a frightening mental image in his mind, one that he continued to repeat over and over again.

THE HEALING BEGINS

It would fall to Dogtown trainer John Garcia to begin the work of helping Rush overcome the psychic damage done by the war. "In the past I have worked more with high-energy, outgoing dogs, because their energy meshes with mine," said John, who is 27, laughs frequently, and has a jolly moon-shaped face and a tiny goatee. "So shy dogs like Rush are a challenge for me."

From the beginning, Rush faced some very high hurdles, as seemingly ordinary circumstances could throw him off. His first visit to the staff room—just an ordinary room filled with tables and chairs—paralyzed Rush with fear. As John attempted to lead him into the room, Rush balked at merely crossing the threshold. He tucked his tail low and turned from the door. Everything about entering the room filled Rush with dread. But by gradually, gently coaxing him into the room, and staying inside it only briefly, John began helping Rush take the first tentative steps into a new world of fearlessness. Every step was rewarded with either a chicken treat, a scratch behind the ears, or some other positive payoff that would encourage Rush to keep moving in the right direction._

Being hit by a car is perhaps the most common injury a dog will face resulting in a leg needing to be amputated. Cancer, tumors, infections, and other diseases may account for most of the remainder.

John felt that Rush may have had a bad experience with a closed room—perhaps in the clinic where his leg was amputated—so he wanted to make sure that the dog understood that when he went into a room, he wasn't going to be poked with needles or hurt in any way.

On the other hand, John's theory may not have been true at all. It was equally possible that Rush simply had little or no experience of being inside a closed room, and that's why he was frightened. Or perhaps it was something else entirely. As is so common when trainers try to penetrate the mysteries

During his time in Beirut, Rush's left front leg had to be amputated due to injury. He gets around quite easily on three legs.

of an animal's mind, John simply had to take an educated guess about what was causing Rush to behave the way he was, and work with that.

Traumatized dogs often resist being touched at all, so John started with simple handling exercises in order to build trust. He also began feeding Rush by hand. He was extremely careful to avoid doing anything that might provoke the dog to bite.

"When you're working with dogs like this, safety is the number one priority," John explained. "But it's not only the safety of the person, it's the behavior of the dog. If he bites me, not only will I be hurt, but he will be hurt also, because he is practicing that behavior. And the more he practices that behavior, the better he's going to get at it. The less he practices it, the more likely that behavior is to go extinct."

One day Rush actually did bite John, though not severely. While John was working with him, Rush nipped him, apparently because he was spooked and not so much because he was intending to attack. Nevertheless, it was worrisome, and meant that for the time being Rush would wear a red collar (meaning that only staff members were allowed to handle him).

Gradually, John began to notice new behaviors, signaled in dog body language, that suggested an increasing sense of relaxation. Once or twice he saw Rush yawn deeply. He saw him lower his chest almost to the ground while raising his tail and hindquarters toward the ceiling in a luxurious, full-body stretch. When he began to relax, it was evident what a beautiful dog Rush was, graceful, long, and lean. German shepherds, and shepherd mixes, are beautiful in the same way that wolves are beautiful, with a look that combines ferocity, wildness, and power.

A POSITIVE APPROACH

Next John turned his attention to grooming, as a way of desensitizing Rush to being touched and handled, and as a way of further building trust. At first, Rush was a "bit skittish," John said. "He wasn't really thrilled by the idea that I was touching him on random parts of his body with a brush—a strange, scary object—in my hand. When you're working with a dog with fear-based behavior, if you pick something up, that is potentially threatening to that dog. If I pick up a poop scooper, if I have a brush in my hand, that's something that a dog may consider very threatening.

"So initially working with Rush with a brush was, uh, 'very interesting' because I wasn't sure exactly how he was going to react to it. He definitely needed to be groomed, but at the same time, I didn't want to just start brushing him all over the place and have him have a bad reaction. So we had to work into it."

John began using the training technique called targeting. If Rush touched something, or investigated something, he got a treat or a reward for it. So when Rush started targeting the brush, getting familiar with something that had been strange and threatening, he got a reward. Then, very briefly, John began using the brush on Rush's back. Each time, the experience was nonthreatening and positively rewarded, so that over time John actually began grooming Rush's tawny coat.

"I think he understood right away what we were doing, and pretty soon I was able to just brush him all the way," John said.

In the coming weeks, John began teaching Rush to sit, stay, and come. Rush was bright and attentive, and learned quickly. As the two continued their partnership, the bonds of trust and affection began to blossom and grow. One day when John, seated on the grass, gave Rush the command to come, Rush rushed up and snuggled John, rolling forward so that his chest and face were in John's lap. John petted and scratched him affectionately; when John stood up and walked away, Rush came pedaling after him with his curious three-legged gallop.

The original Rin Tin Tin amazed audiences in 1922 with his ability to jump—one vertical leap measured 11 feet, 9 inches.

The next step was to help Rush overcome his fear of sharp noises. John loaded the big, tawny dog into his truck and took him down to Kanab Creek, a stream not far from the sanctuary. The creek had steep, sandy banks. It was peaceful and quiet, and nobody was around. The only sounds were of distant birdsong, wind in the trees, and the drowsy slap of water on water. It was, in other words, the polar opposite of war.

John took off his socks and shoes and led Rush down to the water, holding his leash in one hand. John waded in to just below his knees. When John held out a treat, Rush, his body tipped at a steep angle up the sandy bank, gingerly took a few steps into the water to get the treat. Rush kept flicking a wary glance around, tongue hanging out, as though scanning the landscape for enemies.

John moved slowly and deliberately into the mental war zone that Rush had brought with him to the river. John started by snapping off small twigs, which made a barely perceptible popping sound. When Rush did not startle, John rewarded him with a treat and a back scratch. "Good boy, Rush! Good, Rushy-boy!"

Then John started snapping off bigger sticks, then small branches. Then he smacked the water with a stick. Each time, when Rush did not startle, or startled only slightly, John gave him a treat. Next John got out of the water to take Rush on a little walk through the woods.

"I want to go into a dense area where there a lot of sticks, a lot of snapping sounds," he explained. "All the birds and wildlife and weird sounds in the woods—all these things are very strange to him. I can't stress that enough."

Once or twice, Rush flinched at the myriad sounds around him, but he did not lash out or bite. Even hobbling down the woodsy path with that odd, bouncing, three-legged gait, Rush's wolfish beauty was evident. He kept looking back at John with those alert black eyes, his eyebrows pinching up quizzically. He was always glancing around restlessly, like a wild animal, never tranquil, never at rest.

"On the trail he did startle a couple of times, and once he looked back at me in that strange way, which is usually the precursor to him biting somebody," John said.

But he didn't bite anybody. And that was very good indeed.

One sunny afternoon John loaded Rush into his pickup truck and drove him a short distance from the sanctuary to a different magical place. It was one of the few places in the dry canyon lands where grass grew, and most dogs loved it there. There were a few picnic benches and a couple of immense cottonwood trees, but no one else was around.

"We're ready to go to the next step, getting him out, exposing him to new things, getting him to look forward to tomorrow," John explained. In Rush's development, the trip represented a significant step forward, because John could now allow him off leash for a taste of genuine freedom. He could romp around, relax, back scratch in the grass—or run off, if he wanted to. "I felt very confident that he would stick around and listen to me," John said.

One of the special things about this place was a concave canyon wall, which added echoes to sounds. When John clapped his hands sharply, the sound echoed back, multiplied. Now he tried clapping in front of the canyon wall, and though the claps came back, seemingly from all directions, Rush did not budge.

"Not one thing could I do that would actually scare him," John said, with obvious pleasure. "I'd clap my hands, I'd cough, I'd do everything possible that he would startle with before, and there was no reaction at all. So not only did he exceed my expectations, but he's created new behaviors on his own that are very socially acceptable. So it's good stuff! It's good to see that improvement.

"We work with a lot of dogs where you don't see improvement very quickly, and it's easy to give up. It's easy not to have the patience to actually work with the dog. So when you work with a dog like Rush, it's very gratifying, because the amount of time versus the behavior he's exhibiting now is amazing—I mean absolutely amazing!"

FINDING A NEW HOME

The ultimate goal of all this training and desensitizing was to make Rush capable of being adopted and living in a home. Rather than being habituated to gunfire and bomb blasts, he needed to grow accustomed to the hubbub of human life—doors opening and closing, people talking and laughing, dishwashers, garage door openers, lawn mowers.

"Building up to this is a long process, but we have the time to do it," John said. "That's our biggest asset here: time."

"Training dogs is very simple, actually," he added. "It's not rocket science. It just takes a lot of patience, a lot of time, and a pocketful of chicken!"

John decided to start taking Rush home to his house, but the first time Rush walked into the living room—a modest, comfortable interior dominated by three dog beds—he hunched over, lowering both his head and tail. He glanced about fearfully, sniffing the carpet as he went. He seemed tentative and on edge. For some reason, he was particularly terrified of the ceiling fan and would go to great lengths to avoid it. For Rush, interior spaces in general seemed far more alarming than the outdoors. But the more times John brought Rush home, the more relaxed he seemed to become.

"He's making huge strides," John said one day, after a few months of visits to his house. "Today he went right up and investigated the dishwasher. He would never, ever have done something like that before—he used to just tremble at the slightest sound."

Rush has made huge strides—learning to relax in many formerly scary situations—because of his work with Dogtown trainer John Garcia.

Here John gently opened and closed a kitchen drawer while simultaneously handing Rush a treat. "Yeah, remember that, Rush?"

Rush snatched the treat, barely flinching at the sound of the drawer snapping shut.

It was the kitchen, with all its clanking, banging, and bubbling, that posed one of the most daunting hurdles for Rush—which was slightly comical, for a dog who had lived through an earthshaking military assault. But for him, cooking dinner could be as scary as the sound of a rocket-propelled grenade. In fact, it took Rush more than a month of visits to John's kitchen at suppertime before he seemed to really relax.

"We tend to forget that some of the things we do on a daily basis can actually produce a lot of very fearful behavior from dogs, especially dogs like Rush," John said. "Cooking food can be very traumatic to some dogs, believe it or not. The boiling water, the getting the noodles out, the pots and pans, the dishwasher. There's all kinds of things that can be really scary. So just the simple fact that I can now make dinner with Rush is huge in his life.

Because of their intelligence and strength, German shepherds are often used as police dogs and service dogs and excel at agility and obedience.

"Not only is it noisy, with clanking pots and pans, but also there are these aromas that are enticing to him. I want him to experience all this stuff, and to experience it in a home environment. After all, you know, if he's adopted, what's going to happen the first night he goes home? People are going to cook food. So I want him to understand that that's a fun thing, not a fearful or threatening thing. It's sort of a group activity that he's a part of."

Rush, who had been lingering around the kitchen during John's little pep talk, spotted a stuffed lion in the next room and went bounding after it eagerly. He returned with the lion in his mouth, dropped it on the floor, and came up to John with his tail wagging, tongue hanging out, his ears pivoting forward and then back again. He seemed calm, alert, and happy.

"I think he is adoptable right now," John said. "This is the best he's done in the whole time he's been at Best Friends. I think somebody could come in and take some time to manage his behaviors and have a really great dog on their hands. I would love to have him in my own home, actually. He's a really awesome dog, man."

"People have a lot to learn from animals," John continued. "That somebody like Rush, with such a traumatic past, can come so far, so quickly—to me it's just inspiring. In this work that we do, it's very easy to lose track and lose hope, to lose patience. But working with a dog like Rush inspires me. I think it inspires all of us."

That inspiration is based on the demonstrated capacity of dogs like Rush to genuinely change, and even to overcome experiences as psychically brutalizing as war. The inspiration is also based on the hope that one day they will find someone, somewhere, who can love them for who they are. In the summer of 2009, that magic day arrived for Rush when a family who'd seen him on *DogTown* just couldn't resist his charms, three legs, PTSD and all. They drove to the sanctuary to take this modern-day Rin Tin Tin home, to a place—and a life—about as far from war as it was possible to get.

Barnum & Sadie

Michelle Besmehn, Dogtown Manager

Before I came to work at Best Friends, I had a very different life. My home was a trailer towed behind a truck. My partner, Bob, was a professional juggler. We traveled around the country performing. At the time, I didn't have any pets. I liked dogs well enough but hadn't yet had the experience of choosing one for myself. It was certainly new and challenging when I finally got my first dogs; two beagles named Barnum and Sadie. These two characters taught me so much about the ups and downs of having a pet. Today I feel I owe them everything.

It all began with a beagle I saw in a pet store. He was a charming little puppy, and I fell in love. But, even back then, buying a dog from a store just felt wrong. Still, I couldn't get the puppy out of my mind, so I decided to look for a beagle to adopt from a shelter. I started combing the newspaper ads in each town I went through, looking for beagles. On a trip through Casper Wyoming, I saw ad in the paper listing the dogs at the local shelter—one was a two-year-old beagle.

When I got to the shelter, I wavered slightly from my beagle mission. This was the first time I had been at an animal shelter and it was a bit overwhelming. When the staff introduced the dog to me, I was a little taken aback. He was mature, not a cute, little puppy like the one in the pet store. They told me he was about two years old (after some time, I realized that he was more likely five or six). His manner was very reserved. When I first met him, he didn't approach me with a greeting (in fact, he was the only dog I met that day who didn't come up to say hello). But it didn't matter: I took him home that night anyway. My first shelter visit had an enormous impact on me. I remember crying when I left the shelter that day with my new beagle. Bob asked, "Why are you crying? You adopted one of the dogs." I cried because of all of the other dogs that I had to leave behind.

The little guy coughed and coughed the whole ride home. When we arrived, the first thing I did was give him a bath. The next thing I did was give him a name: Barnum. Barnum's coughing grew worse, deeper and more honking. I had no idea what could be causing it so I took Barnum to the vet, who diagnosed the problem and prescribed some medication. It turned out to be "kennel cough," a highly contagious and very common illness in shelter dogs. Barnum got better, and we were through our first bump in the road.

But our next bump wasn't very far away. Barnum was a pretty low-key dog most of the time, but he became very anxious when he was left alone for certain periods of time. Barnum had the uncanny ability to differentiate when we were leaving to go to work as opposed to going out for fun. This is a difficult problem, but through trial and error I found solutions. One way was to let Barnum ride along with us. He would sit in the cab of the truck (as long as weather permitted) and wait quietly for us. Barnum loved to travel and felt safe in the truck, so not only did he feel like he had been on an outing but we could leave without worrying that he would destroy something. It was fun for him and he got to go lots of places. Now after having worked with dogs for many years I realize that there are many other things I could have done to help Barnum through his anxiety about being left alone.

Looking back, I could have done so many things differently. I was attracted to beagles without knowing beagle characteristics. I based my decision on an emotional response to seeing an adorable puppy at a pet store. If I had done my homework, I would have been more prepared for the types of behavior and medical challenges I might encounter. Once I got Barnum home, I didn't know how to train him. I was learning as I went. But the amazing thing was that, despite all his imperfections, I fell in love with this squishy-faced beagle. Flaws and all, I adored him. I loved every hair on his body, and Barnum loved me right back. What made our bond even stronger was how he was gracious and loving enough to tolerate all the mistakes I was making.

Barnum had a quirky personality. Other than me, there were very few people that he would give the time of day to, unless there was food involved. Then he demanded that people notice him. Barnum would near your feet, and if you ignored him, he would start making little grunts and woofs. If you still failed to pay attention and no food came his way, then he would bay. Barnum was a true character.

After Barnum had been in my life for about a year, I drove through the same town and dropped by the same shelter where I had adopted him, in search of a companion for Barnum. I adopted a little female beagle, whom I named Sadie. Even though Sadie was going to be a companion for Barnum, I didn't take him with me when I went to select the dog, so I had no idea if Barnum would even like this new dog or if she would like him. I took Sadie home—never realizing that all dogs didn't necessarily get along; I just thought they would meet each other and hit it off right away. What I imagined would be a joyful meeting turned out to be my first lesson in dog introductions. Sadie immediately picked a fight with Barnum.

In the following weeks, the situation didn't improve much, and I was beginning to worry. Sadie started off her first night by sleeping in the middle of our kitchen table. She continued to be hostile toward Barnum (whom I adored) which really upset me. I wasn't sure how I felt about my new dog, she didn't seem to connect with me, and she definitely didn't like Barnum. But, thinking back, I suppose I wasn't being fair to Sadie.

I had just uprooted her from whatever life she had known and placed her in a new situation in a strange place with a strange dog. I had these unrealistic expectations that she would just magically know the rules of our home and fit in perfectly from the start. In reality, Sadie didn't know what was going on, and she didn't have a clue what was expected of her. As we learned together and began to understand each other, Sadie started calming down. The unease I had about my mixed emotions toward this difficult dog went away. I grew to love Sadie, too.

Eventually Sadie became a cuddle bug. She couldn't get physically close enough to me. If she could have, she would have crawled inside of me. There wasn't much room in the trailer, so I didn't often let the dogs sleep in the bed with me. But clever Sadie figured out that if she got in bed before anyone else and fell asleep quickly, she wouldn't be evicted. As she grew more comfortable in our home, Sadie's silly side emerged. A true clown, she would grab a piece of laundry if I was folding clothes and run around, hoping that I would chase her.

The beautiful thing about bringing Sadie into our lives was that it helped me to become a better dog owner. My days of doing everything wrong were behind me. I was learning to work with my dogs in a way that felt comfortable and right to me. I encouraged positive behavior instead of punishing for the bad. Using positive reinforcement helped bring Sadie and Barnum together. Anytime they had a pleasant interaction with each other, I rewarded them. In that way, they began to associate getting along with praise and treats. Soon, they began to enjoy each other's company without needing a reward. They began to play with each other on their walks, and a deep friendship grew between them. All three of us had come a long way to get to this happy stage.

Although Barnum and Sadie brought a lot of happiness and laughter to my life, they were never easy dogs. Food drove them crazy to the point where we had to put all the food in "beagleproof" containers, or they would eat it. I even came home once to an open refrigerator with most of the contents all over the kitchen floor. I never quite figured out how they got it open—but I basically had to beagleproof my living space after that incident.

Both beagles had medical issues. Barnum started having health problems around the time we got Sadie. His eyes started swelling up, and after trips to several different vets, he was finally diagnosed with glaucoma, a condition that would trouble him for the rest of his life. A few years later, the glaucoma eventually took his sight in one eye. Eventually a cataract in his remaining good eye caused him to go completely blind. Sadie started having occasional seizures which is apparently not uncommon for beagles. The seizures were never frequent enough to warrant medication. Sadie seemed to sense the onset of a seizure and would go to a quiet, safe place (a favorite dog bed or next to me on the bed).

As Barnum got older, he developed heart issues that became severe. He got very sick about seven years later. It was then that I learned the hardest lesson of all: knowing when to say goodbye. Letting Barnum go was really difficult, but I could tell his physical problems were taking the joy out of his life. He couldn't get up on my bed anymore, so I slept on the floor with him. When it was time to let him go, it broke my heart. I loved both of my beagles very much, but Barnum was the love of my life. I had a unique bond with him, one that I've not experienced again. He was the most trusting dog I've ever known. Even when he was completely blind, I could run with him on leash and he would just run beside me with abandon, trusting that I would never let him run into anything, that I would never let anything happen to him. As his health continued to deteriorate, I wondered if I had waited too long, if he was suffering too much. But Barnum's eyes met mine before he died, and I knew that he forgave me for every mistake that I had ever made with him. I was still mourning Barnum when, just a few months later, a sudden illness took Sadie. I often think back and wonder if the bond that they had formed contributed somehow to Sadie's untimely death. .

Barnum and Sadie taught me so much. They made me appreciate how important it is not to give up, which is something I carry with me to this day on the job at Dogtown. Barnum and Sadie were instrumental in helping me see dogs as individuals who should be appreciated for their particular characteristics, both good and bad. They taught me that dogs are not preprogrammed with a set of rules. It is our responsibility to work with our dogs to establish rules, to communicate them to the dogs in a way they understand, and then to help the dogs succeed in following them. They taught me that sometimes, the bonding process isn't immediate and can take some time. It took some time for me to fall in love with Sadie, which seemed terrible when I went through it. I see the same reactions in adopters at Dogtown. They want so much to form an instant bond, but sometimes it takes time and effort. My experience with Sadie helps me explain this to people, so that they don't feel bad if the thunderbolt doesn't hit them right away.

Although I love beagles, I have never adopted another one because I am afraid that I will look too hard for Barnum and Sadie in them and not appreciate them enough for who they are as individuals. These two dogs inspired my passion for adoption and for giving unwanted animals a place of their own and a chance to find their perfect match. Barnum and Sadie set me on the path to Dogtown, and it's something for which I thank them every day.

Wise old Bruno's thick red coat needed a trim when he arrived at Dogtown.

Bruno: Last Days of the Cinnamon Bear

Imagine an elderly man in a flimsy hospital gown, blind in one eye, nauseated, disoriented, and with multiple health problems, who cannot care for himself or even speak, being dropped on a stranger's doorstep.

That's what happened to Bruno.

He was an old dog, down on his luck, tottering on his feet, drooling, dizzy, and nearly deaf. His problems had become so overwhelming, in fact, that his owner had simply dropped him off at an animal shelter in East Los Angeles. Perhaps daunted by the potential vet bill, or the demands of taking care of Bruno during his declining years, the owner had essentially dumped him out the back of the car—knowing full well, no doubt, that at the shelter a dog like Bruno would probably have only a few days until he was euthanized.

Yet as broken down as he was, Bruno could still break some hearts. He looked like a cinnamon-colored teddy bear, with eyes so dark they were nearly black; a shiny black nose; and short, furry, forward-facing ears. An extravagant ruff of red fur encircling his face like a mane gave him, at the same time, the look of a lion. His head was so richly mantled in reddish fur that it seemed to nearly bury his black eyes, and he had a line of darker fur down the center of his forehead, the way a lion does. But his demeanor was almost completely devoid of a lion's ferocity; despite his discomfort, Bruno's mood was mellow. In fact, when he peered out of his cage at the shelter, with his head tilted quizzically to one side, he looked about as forbidding as a plush toy.

It was hard to believe, seeing him in such sad circumstances, that Bruno came from an illustrious lineage going back 4,000 years. He was a chow chow, or chow; one of the oldest dog breeds, it is thought to have been developed by the fierce horse warriors of Mongolia, and later spread to China as early as 150 B.C. With their distinctive black mouth and tongue, chows are bundled in such resplendent fur that the Chinese had a name for them that meant "Puffy-Lion Dogs."

But Bruno looked less like a puffy lion and more like a sad, sick old man when Best Friends volunteer Nadine Goodreau spotted him at the shelter. Nadine, who ran the Los Angeles chapter of the volunteer group called Best Friends Brigade, worked on the front lines trying to rescue the neediest animals from California shelters. And she was immediately touched by this gentle old dog, with his serene, lion-bear face, peering out of a cage. She could tell that Bruno required serious medical attention and vowed to find him a better life.

To start, Bruno's coat needed some care. The old dog's fur was so hopelessly matted that it took Nadine and a couple of helpers two days (three hours each day) to shear it down to the skin in a "lion cut," leaving red fur on his legs and head. They even shaved his tail except for a comical little pom-pom of fur at the end. His cinnamon fur would grow back rich and radiant, but for now he just looked amusing: His endearing face, still regally adorned with fur, looked almost as though it were disconnected from the rest of his body, which was naked as a freshly shorn sheep.

Bruno stayed calm and patient while Nadine administered his beauty treatment, but "while he was at the shelter I believe he lost his spirit," she later wrote. "From that point on, I was on a mission to not have him die at the shelter. Bruno needed a better ending to his life, and I was going to see to that."

Nadine "networked like crazy" but could find no rescuer or private party that would take the old chow. Finally, she decided to "bail him out" of the shelter and take him to her home temporarily, until she could find a better place for him. She was amazed to discover that good-natured

Surrendered to an East Los Angeles animal shelter, Bruno made his way to Dogtown via the L.A. chapter of the Best Friends Brigade.

Bruno was calmly tolerant of all her dogs and cats. After three weeks, she got the good news that Old Friends—the part of Dogtown reserved for elderly dogs—had a spot for Bruno, so she and a couple of friends made the eight-hour drive from Los Angeles to Kanab, Utah, to bring Bruno to the sanctuary and the medical care he needed and deserved.

When she arrived at Dogtown after the exhausting drive from L.A., vet tech Jeff Popowich helped her take Bruno out of his crate in the back of her SUV. She told Jeff everything she could about him, but admittedly, she didn't know much about Bruno's history—his owner had given nothing to the shelter when Bruno was surrendered. His medical history, from his present condition to vaccinations, was a mystery. "He's extremely mellow," she told Jeff. "And he's OK with dogs or with cats."

Nadine, a striking middle-aged woman with dark hair and hoop earrings, wearing blue jeans and a denim jacket, mentioned that Bruno hadn't urinated during the whole eight-hour trip from California. That was worrisome: Jeff knew from experience that it can be a bad sign if a dog urinates very infrequently.

Nadine had put Bruno on a leash and was leading him up the steps into the clinic when he stumbled on the stairs. Jeff, whose job was to make the first, preliminary assessments of animals being admitted to Dogtown, grew more concerned. Bruno's weakness and disorientation were obvious, but the cause of these symptoms could be ascertained only by Dr. Mike, Dogtown's medical director and head vet.

"LET'S LOOK AT YOUR GRILL, KID"

You didn't have to spend much time with Bruno to realize something was seriously wrong. When Jeff got him into an examining room at the clinic, he immediately noticed that the touching way Bruno tilted his head to one side was permanent. His head was always leaning to the left. Bruno constantly drooled; a silvery necklace of spittle was permanently suspended from the left side of his mouth. Bruno also seemed perpetually off-balance, as if he'd just kicked back a martini or two. Jeff knew that these could be signs of a serious medical issue for an elderly dog.

Jeff lifted Bruno up onto an examining table for his assessment. "Let's look at your grill, kid," Jeff said, taking Bruno's muzzle in his hand and gently lifting his jaws apart. "Can we open up?"

Bruno good-naturedly opened his black mouth and gave Jeff a putrid blast of bad breath. Then he sneezed all over him.

"Whoa! Thanks, man!"

Jeff noted on his exam charts that Bruno's breath was horrible, and that he seemed to have some lip fold pyoderma, meaning that where drool had been seeping into the fur below his mouth it had become infected and malodorous.

Still, "when you get a dog like Bruno, you're not sure if he's happy or if he's miserable," Jeff said. For one thing, his black eyes were obscured by his dense mat of paprika-colored fur. For another, chows are by nature reserved, regal, and difficult to read, like cats—or, perhaps, like lions. Bruno's calm state didn't reveal much about his physical state: "Bruno's a blank slate—we don't know the pain he's in," Jeff said. "And it's not like we can talk to him."

Jeff, big as a barn door and sporting a buzz cut, a goatee, a couple of days' growth of stubble, and a gold earring, likes to park his sunglasses on top of his head, where there's an old, ugly scar showing through just above the hairline. He has the intimidating physical presence of a bouncer or a bull rider; the scar suggests a story you'd probably rather not know about.

Two common symptoms for dogs with painful mouth disease are excessive drooling and bad breath.

"I've got a chow named Fuzzy at home, and I know how they can mask things like pain or discomfort," said Jeff, who looks as if he might do the same. "Fuzzy's nickname is Mr. Personality, because you can never tell what he's thinking."

Even so, Jeff said, "Sometimes you make connections with dogs and it's just instant. I don't know if it's because Bruno is so mellow, or the little head tilt, or the little drool that comes down, but there's something about that dog that you just say, 'You know what, you need a treat and a pet on the head. You need something.' It's just easy to give to a dog like that.

"Bruno is a dog that, one, he has a pulse, and two, he's cute as hell. So, I mean, from there you can't go wrong. Most people would say, 'There's a dog knocking on death's door.' But we don't see it that way here. Bruno's a dog that with a little TLC, he can be bouncing around. He can get a second chance at life."

A VISIT WITH DR. MIKE

After the assessment, Jeff turned to getting Bruno cleaned up. He had to give the old boy a bath, clean out his ears, and put some medication in his left eye, which looked caved in and was oozing. He noted that the hair around his lips needed to be shaved and treated with medication to help clear up the festering infection and the bad case of dog breath.

Next up was a medical exam from Dr. Mike. But before Bruno even got a chance to see the vet, the old dog fell on his side and seized for a few seconds. Jeff saw the seizure. "It wasn't a very long one, not too severe, but it was still frightening," Jeff said. "Seizures are unpleasant to watch, but really you just have to let the dogs ride it out and hope they don't bang their head on the floor or hurt themselves. But it can be scary if you've never seen it before."

For Bruno, it was one more ominous warning that something was seriously wrong.

When Jeff took Bruno to the clinic to see Dr. Mike, he told him about the seizure, and about Bruno's disorientation, and said that he seemed more comfortable turning to the left than the right. His head tilted left, and he drooled to the left.

Dr. Mike lifted the old chow onto an exam table and peered into first one ear, then the other. "In a younger dog, we would suspect infection was interfering with his balance," Dr. Mike explained. "In older dogs, it could be from a vascular accident in the brain similar to a stroke, an inner ear infection, a tumor, or some other unknown reason (often called old dog vestibular disease)."

Bruno, as he did in most situations, good-naturedly tolerated the ear exam. But at the end of it, he furiously shook his head, like a big wet bear. "Good boy," Dr. Mike said. "Not too bad in the ears."

Dr. Mike pushed back the ruff of red fur to peer into Bruno's left eye and quickly realized that he was blind in that eye. "That could partly explain what's going on," Dr. Mike said. But he suspected something deeper was going on in Bruno's brain. A larger neurological condition was more likely the root of Bruno's problems.

He lifted Bruno down onto the examining room floor to check his balance. Bruno woozily thumped into a door, like the last drunk to leave the bar. Dr. Mike felt his old, skinny hips and found a bit of arthritis there. Musculoskeletal problems are common in older dogs, he explained, but the lack of balance suggested something more serious.

He lifted Bruno back onto the exam table and arranged Bruno's feet so that they were crossed at the ankles, an entirely unnatural position. A healthy dog would quickly right the legs, but Bruno just stood there dumbly, almost as if he were not aware of his legs at all. "All the symptoms," Dr. Mike said, "point to brain disease."

That was very bad.

Still, Jeff Popowich remained buoyantly optimistic.

"We take in dogs that look worse than Bruno on a regular basis here—animals hit by cars, or with really terrible illnesses," he said. "But even if it's a brain tumor, we can still give him some quality of life. Here at the sanctuary it's quality over quantity, and Bruno is definitely a dog that has a potential to have a quality life, even if it turns out to be a short one."

Dr. Mike put Bruno in a kennel in the clinic to keep him comfortable and observe him.

The next day, Bruno was vomiting and seemed to be having a hard time breathing. X-rays of Bruno's abdomen showed that he had aspiration pneumonia (inflammation of the lungs caused by inhaling vomit). He also had a condition called megaesophagus (abnormal distension of the lower part of the esophagus because of problems passing food into the stomach). Dr. Mike put him on IV fluids and antibiotics, and placed Bruno's food so he could eat standing up. Within two days, Bruno's mood had improved, his energy level was up, and his food was staying down. X-rays taken three days and then ten days later showed that Bruno's lungs and esophagus had improved greatly, and his balance and breathing seemed better, too.

THE BRUNO FAN CLUB

"Bruno's temperament seems quite sweet," Dr. Mike concluded in notes he posted on the Guardian Angel website, an online forum for special-needs animals and the people who are touched by them. "I believe we have done all we can medically for him at this time. Bruno is an older dog . . . who I hope that we can make comfortable for the remainder of his life."

A photograph of the lovable cinnamon bear was also posted, along with Dr. Mike's medical notes. It brought an outpouring of sympathy from those who visited the site:

Bruno is so stoic and dignified . . . yet he has the most endearing overbite. A major mush-magnet for me! Bruno is irresistible.

I find it upsetting that the "owner" just gave him away after such obvious neglect. But at least they had the good sense to take him to a shelter that could help instead of just letting him die . . . I, for one, can't get enough of this sweet old guy!

OMG, that is a picture to shred the heart. I cannot imagine that poor old boy out alone on his own, trying to survive and being so sick. Rest easy, Bruno. You're home, finally, with the best pack that exists.

Dr. Mike gave the go-ahead to have Bruno placed in the "geriatric run" at Old Friends. Bruno would share an "apartment" with the five oldest dogs at Dogtown, all of whom shared his mellow temperament. Each apartment had a comfortable indoor room, a dog door, and an outdoor run. For now, it was home at last.

Dog trainer Ann Allums helped introduce Bruno to each of his new geriatric roommates one by one, so he wouldn't be overwhelmed. The first dog, Kanani, was a splendid female husky. As soon as he saw her, Bruno instantly perked up, wagging his tail and following her around like a small boy with his first crush. But Kanani haughtily spurned Bruno's advances.

Next he was introduced to a spunky little male terrier named Mr. Pepper. Jeff, who was watching all this, noticed that Bruno got "all stiff and macho, like a guy walking into a bar" when he was introduced to Mr. Pepper and the other male dogs—a little cocker spaniel named Quazimoto, and Clarence, a blind border collie. But when he was introduced to another female, a sweet-tempered shepherd mix named Chili, Bruno went on high alert again, wagging his tail and bird-dogging her around the kennel.

Sly old Bruno, it turned out, was quite the ladies' man.

When Ann posted a note about Bruno's move to Old Friends on the Guardian Angel site, more adoring fan mail came pouring in. Bruno was becoming a kind of canine rock star:

Bless you sweet Bruno!!! Perhaps Kanani and Chili are just being coy. In any case, I'm glad to hear that something or someone put a spring in Bruno's step, and he's making friends in his new digs. Thank you to everyone for taking care of this dear boy!

Hey, old-timer! You still have the stuff, huh? Enjoy your new friends and the BF geri-spa routine. I see you got your haircut already . . .

It's so nice to hear that Bruno is feeling better and has his own run now. I bet once that beautiful paprika-colored fur starts to grow back in, the ladies will notice him more . . .

Jeff, keeping track of the old boy each day, noticed glimmers of improvement. His back end, once listing like an old barn, seemed a little bit more stable. The pneumonia had cleared up. And he was peeing more regularly—"and everybody is kind of happy about that one."

Ever the optimist, Jeff kept hoping someone would appear at the sanctuary, or see his picture online, and decide to adopt him.

"Now that he's out there in his run, all it takes is one person to come by and fall in love with that guy and take him home," he said. "It's got to be somebody who is OK having to say goodbye in a year or two, but there are people like that all over the place. I haven't seen anything about Bruno that says he's not adoptable. Of course, I've never met a dog that's not adoptable either. I'm still optimistic—I always am. I think he's gonna be OK."

A TURN FOR THE WORSE

Unfortunately, Bruno's health began failing again not long after he arrived at Old Friends. The vomiting returned. Increasingly dizzy and disoriented,

Bruno began to bump into things more than he had before. Dr. Mike decided to move him back to the clinic, for observation.

Even so, in the mere week he had been at Old Friends, the sweet-faced cinnamon bear charmed a couple of volunteers to the point that they put in an adoption application for him. They wanted to take him back to their home in Montana. Now the question became: Could Bruno be stabilized enough to make the ten-hour trip to Montana? Dr. Mike had some serious concerns about the long car trip; it seemed to him that the trip from Los Angeles to Kanab might have played a part in Bruno's decline upon his arrival at Best Friends. The vet wanted to play it safe and monitor Bruno's condition for a little longer before discharging him into a new home.

If a dog is having a seizure, it is unnecessary to place an object between the dog's teeth, or to pull out the dog's tongue—dogs are unable to swallow their tongue.

At the clinic, Bruno's condition seemed to improve for a couple of days. Still, "improvement" was relative; he threw up only twice in a couple of days, which was less frequently than he had been throwing up at Old Friends. Then he genuinely did seem to improve, going four straight days without vomiting.

But the hard truth was that Dr. Mike and the veterinary staff had been unable to precisely pinpoint the cause of Bruno's problems, so they had no way of knowing whether Bruno would live a few more hours, a few more days, or a few more years. Dr. Mike believed Bruno was now probably a good candidate for animal hospice—caregivers who would simply ease his pain as he gradually slipped away.

When Dr. Mike posted Bruno's medical updates online, Bruno's fan club enthusiasm seemed to be swelling into a standing ovation:

Sweet Bruno, if only you knew how many people are cheering and praying for you!

Bruno—I'm so sorry to hear of this setback. I hope all goes well for you, and you regain good health. Bruno seems so sweet. Hang in there, big guy!

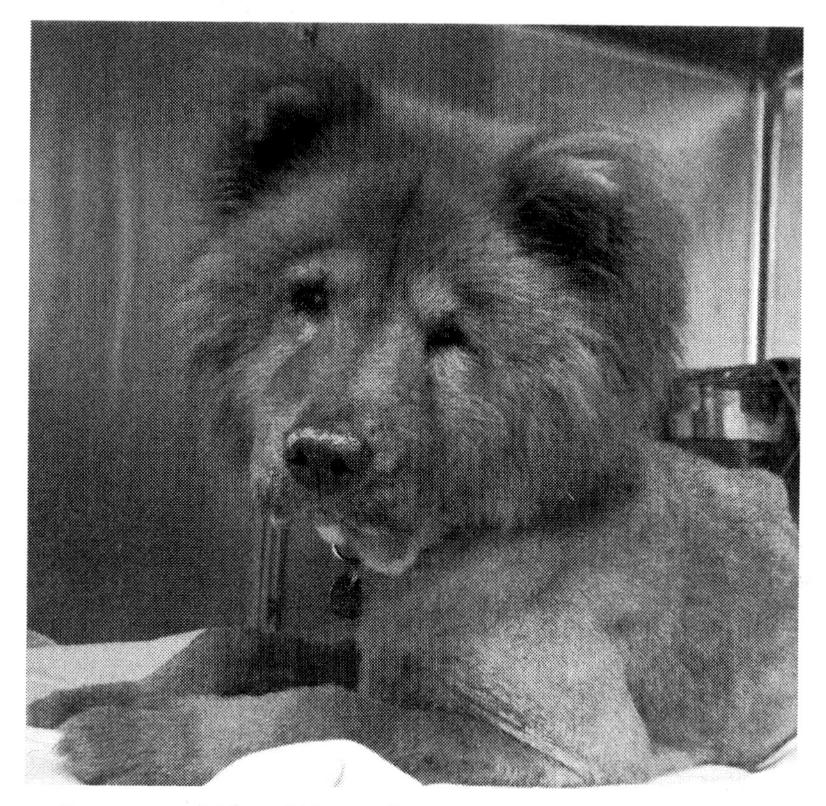

Bruno enjoyed life at Old Friends, Dogtown's unit for senior residents, until his health worsened, and he returned to the clinic.

Still, as Dr. Mike's wife, Elissa Jones, wrote in a posting on the site, "Life in a cage isn't much of a life, so we decided to take him home and foster him. Doing that would get Bruno out of a cage and out of the heat. We'd be able to make him as comfortable as possible and he'd have a veterinarian at his beck and call." Part of their decision was to assess how Bruno would do in a home and see if the trip to Montana was really an option for him.

The Jones-Dix household was essentially two human lives entangled in the lives of a menagerie of dogs and cats rescued from various heart-rending plights. "How'd we get all these animals? Just happened," Dr. Mike said, looking around at the cats walking across the counter.

Since Nadine had already told them that Bruno got along well with dogs and cats, Dr. Mike and Elissa were not surprised when the big puffy-lion dog effortlessly slipped into their household. They set Bruno up in his own room with a dog bed and a baby gate, so he could have a quiet space away from the rest of the dogs.

But the 20-mile drive to their house from Best Friends, in the desert heat of June, was difficult for him, and when he got "home" he paced restlessly, bumping into things, and breathing with difficulty. Both Dr. Mike and Elissa worried that Bruno might not even make it through the night. Mike put food in a dish elevated a few feet off the floor, but Bruno struggled to eat, a long necklace of drool suspended from his mouth. Most of the time he lay on his bed, a big red half-naked lion-bear, not lifting his head or moving much. It was becoming clear that the ten-hour drive to Montana would probably be too stressful to Bruno's system.

The next day, Bruno seemed to be feeling better. Elissa gave him tiny meals, three times a day, with only a little water, to keep him from vomiting. He was now on multiple medications—anti-inflammatory steroids and drugs for motion sickness and pain. Elissa and Dr. Mike took him for a very short walk and he seemed to perk up a bit.

Still, Dr. Mike said, "We worry if this is the right thing for him—or are we just keeping him alive for the sake of keeping him alive?"

The hours passed. Bruno seemed to get more and more disoriented, circling aimlessly, occasionally vomiting. The majesty of his bloodline, the Mongolian horse warriors and Chinese noblemen who loved and revered his breed—all that seemed a million miles away. Now the sweet lion-dog was just a shadow of himself, thumping into a door and standing there dumbly, drooling.

Then there was a surprise. Elissa wrote in her post that Bruno "tricked us into having three really great days." He stopped vomiting. He stopped bumping into things. He approached Elissa and Dr. Mike to be petted, and seemed to really see them. True to form, Bruno even "developed a crush on our elderly female Dalmatian and we took them on several short walks together where they took turns sniffing each other and marking things." Elissa and Mike began to think Bruno might actually recover, that he might be able to make the trip to a new home in Montana. If that didn't work, they decided that their home would be Bruno's last one, "and we told him that, in case he understood," Elissa wrote.

Bruno didn't do well in the heat, so they limited his walks to nighttime or early morning, when it was coolest. Then, on Friday night, Bruno inexplicably began to spiral downward again. He seemed to be having trouble breathing. He also began to show possible signs of bloat, an abdominal swelling caused by gas or swallowing air. Dr. Mike knew that the only real treatment for bloat was abdominal surgery, and there was no way old Bruno could withstand that. Dr. Mike made the merciful decision to put Bruno to sleep to end to the old boy's suffering.

"We spent time with Bruno petting him and talking to him and assuring him that he was loved as we helped him to ease over the Rainbow Bridge," Elissa wrote, referring to a place "just this side of heaven" where companion animals are said to go when they die, and across which they can communicate with their human companions.

Watching a dog walk into a completely dark room that he is unfamiliar with, and then repeating the test with the lights on, is a good method to determine a dog's vision. A blind dog will act the same way in both instances.

And not long after that, the old cinnamon bear shed his tumble-down body and all his pain and passed over the Rainbow Bridge. He stopped breathing. His body grew still. It was over.

"I don't know what Bruno was like in his prime," Elissa wrote. "I don't know what happened in Bruno's life before he ended up on death row in a shelter for being old and 'unwanted.' I do know that blessings surrounded him after that. . . . At the end, he was in a home, surrounded by love. Those who spent time with Bruno felt blessed to be with him. I know that I did."

Elissa's account of Bruno's last days brought an outpouring of grief from those who had followed his story on the Guardian Angel website:

Elissa and Mike, thank you so much for all you did for this lovely old boy. I don't usually weep when I read these notices, but this story made me tear up—for Bruno, and for all those who tried so hard to bring him back . . .

My heart aches so much . . . I have a friend who is a doctor for terminally ill cancer patients and he told me once that they will be very sick and unaware, and then have one or two very lucid very good days before passing on. Seems it is a way to say goodbye . . .

They say when you find a copper penny it will bring you luck. I just know that everyone who met this bright, copper-colored dog feels lucky to have known Bruno, even only for a brief time. God's speed, Bruno!

SAYING GOODBYE TO BRUNO

Lenny Domyan is the caretaker at Angels Rest cemetery, just a short distance up the road from Dogtown. With brawny forearms illustrated with tattoos, Lenny is an older man whose white goatee gives him a slight resemblance to Colonel Sanders. He has done many "placements" at Angels Rest, where there are thousands of other graves, all marked with the names of cherished companion animals—dogs, cats, horses, birds, rabbits, potbellied pigs: Magic, Apollo, Eclipse, Bobby Magee, Woofie, Two Bits, Goldfinger, Hercules, Cookie Monster, Rocco, Rex, Chewbacca, Lilly, Tico, Sundance . . .

Lenny lowered the tailgate of his pickup truck, where Bruno's body lay, covered with a white sheet. He ran his hand softly over the startlingly inert form.

"Not to worry now, honey," he said, in a surprisingly gentle voice for such a big man. With a black magic marker, he wrote "BRUNO" on Bruno's green, now empty collar. Then he picked up Bruno's shroud-wrapped body, holding it against his chest, and walked toward the freshly dug grave. The body, and the grave, were the size of a child.

Bruno's green collar rests on his marker at Angel's Rest, the animal cemetery at Best Friends.

Jeff Popowich stood watching, along with a half dozen others, as Lenny and another man lowered the old puffy-lion dog into a small grave cut into the rocky ground.

"There you go, friend," Lenny murmured. Then he and the other man silently shoveled dirt and rocks into the hole. "I'm going to do a short reading and then, as we normally do, you can pay your final respects," Lenny said quietly to the gathered mourners.

He read out of a little book: "He demanded so little of us—fresh water, food, a patch of sunlight . . . and he gave so much in return—uncritical, undemanding, unlimited affection . . . he will always be present in our hearts . . ."

Elissa Jones stood at the grave's edge unsuccessfully fighting back tears.

Lenny lay a flat, square, reddish stone on the grave, then set Bruno's green collar on top. One by one, people stepped up and put little polished stones on the gravestone, to honor Bruno's memory. A light wind stirred the dozens of wind chimes hanging in the trees, like an aeolian harp being stirred by the wind's ghostly hand.

"I've never done a placement at Angels Rest where the wind chimes failed to sound," Lenny said. "The chimes are a signal from Bruno that everything is OK. Bruno is over at Rainbow Bridge, whole and happy and healthy and runnin' around—and in need of nothin.' See, when he's over there runnin' around, he kicks up the wind and the wind finds its way over Rainbow Bridge, down into Angel Canyon and strikes a chime in Angels Rest, so we know he's OK."

Angels Rest, the pet cemetery at Dogtown, has been in operation for more than 20 years and provides a final resting place for thousands of former Best Friends animals and pets.

The Rainbow Bridge could be thought of as a self-serving fantasy, a sort of never-never land where cherished animals are supposed to go after they die. But out here in this ancient, celestial place, where the wind chimes keep resonating long after they've been struck, and the desert sky is limitless as the deeps of space, the Rainbow Bridge seems just as real, just as plausible, as anything else. Because the bridges that form between humans and their companion animals—including all the cripples, strays, refugees, and left-behinds like Bruno, the beloved cinnamon bear—are as enduring as the red-rock canyon walls.

Engraved on a stone at the entrance to Angels Rest is a poem meant to comfort the grieving human heart when those bonds are broken. But it also suggests that mortality, human or animal, is really just an illusion—a sort of Rainbow Bridge to the beyond:

"Don't Weep for Me"
Do not stand at my grave and weep
I am not there

I do not sleep.
I am a thousand winds that blow
I am the diamond glints on snow
I am the sunlight on ripened grain
I am the gentle autumn rain.

When you awaken in the morning's hush
I am the swift uplifting rush
Of quiet birds in circled flight.
I am the soft star that shines at night.
Do not stand by my grave and cry.
I am not there.
I did not die.

Kaiser and Sherman

Jeff Popowich, Animal Care Operations Manager

As I sit here writing, my dog Kaiser is sleeping on my couch. He's peaceful and calm, totally oblivious of me, my other dog, Sherman, or anything else going on. I've known Kaiser since around the time I started working for Best Friends. Back then, he was a very different dog. Today, Kaiser is a big, healthy, brown-and-white shepherd, but in his early days at Best Friends, he was underweight and small. Kaiser was also such a nervous wreck that he could barely look at me if I was in the room. Looking at him now, so completely comfortable and so relaxed, it's hard to imagine that he was ever that shy and closed off. But thanks to the help of big brother Sherman, Kaiser has become the easygoing guy he is today. Through Kaiser I learned to work for his trust and friendship, a reward I now get to enjoy every day.

In September 2003, I applied for a job at Best Friends. They were interested in hiring me, but before I could be brought on officially, I had to go through a two-week on-site evaluation. It essentially was a trial run where the staff at Best Friends and I could see if I would be a good fit for the society. They started me off in the admissions area at Dogtown, where all the new dogs are brought in. They wait in admissions while their medical work is being done; it's a good time to get to know the dogs behaviorally, too. It seemed like a good place to start: the new guy working with the new dogs.

Kaiser came to Best Friends at about the same time. He had been sheltered by the San Francisco Society for the Prevention of Cruelty to Animals. They "traded" him to Dogtown because they were having problems finding a home for him and thought Best Friends might have a better shot at placing him in a home. The staff in San Francisco told us that Kaiser was very shy around people. He would sit all the way in the back of his cage and bark a "Stay away!" warning at anyone who walked by. Because of his nerves, Kaiser didn't each very much, and it showed. When he arrived at Best Friends, I could tell by looking at him that he was very underweight (his medical exam later showed that he weighed 20 pounds less than normal for a dog of his size).

Kaiser wound up being one of the dogs I worked with during my two-week trial run, but it was not love at first sight for the two of us. He was very fearful of me and wouldn't come near; when I entered his enclosure, he would quickly move to the point farthest away from where I was standing. I tried not to take it personally. I liked Kaiser and wanted to help him, so I kept an eye out for ways that I could show him my friendship without scaring him.

Every morning we had to leash the dogs up to take them out to the runs, and then bring them back in the building at night. The walks didn't seem to be helping Kaiser, who seemed more fearful when it was time to go for a walk. Trying to leash up a nervous dog can often make the situation worse, as the animal will try to stay away from you, which is just what happened with Kaiser in the beginning. But one morning, I brought Sherman with me, and that changed everything.

Sherman, a big black-and-tan shepherd mix who was rescued from the streets of Tijuana, went to work with me every day (which is a great perk at Dogtown, by the way). He had a few of his own responsibilities, mostly accompanying me and sanctuary dogs on walks and finding sticks along the way to carry around. It was a pretty good life for him, and he seemed to enjoy it.

Sometimes we don't know much about the dogs who come to the sanctuary, so we have to do our own evaluations to assess their personalities, learn their likes and dislikes, and get a sense of who they are while we work with them and try to find them homes. During my tryout at Dogtown, Sherman came in handy when we evaluated how new admissions got along with other dogs. He's a pretty mellow dog who likes other dogs; we didn't need to be too concerned with his showing aggressive behvaior toward new dogs. It was always good to have him around to see how they would interact with him.

With Kaiser, introducing Sherman to him turned out to be the best way to start building a relationship among the three of us. The morning I brought Sherman into Kaiser's run brought about an amazing transformation in Kaiser's behavior. The instant Sherman came in, the two of them started playing together. They were running, wrestling, and jumping in and out of the kiddie pool. It was a puppyish side of Kaiser I hadn't seen before.

Kaiser didn't even seem to mind my presence in there with the two of them as they romped and splashed. He didn't retreat into the farthest corner when he noticed me. Even the leash ceased to be as scary to him; leashing Kaiser up for his walk became a whole lot easier. It was the start of beautiful friendship: Sherman helped Kaiser to overcome his fear and that helped build a better relationship between Kaiser and me.

After my two-week evaluation ended, I was offered the job at Dogtown. It was an exciting moment, but before I could start full-time, I needed to take about a month to wrap up things in my old life. When I came back to start my job, I went back to work in the admissions area, taking care of the new arrivals. While I was gone, Kaiser had moved on and been placed in a run with other dogs. At first, it went well for him, and he seemed to get along well with his runmates.

But in that month's time, Kaiser seemed to have forgotten Sherman and me; when I went to visit him, it was like starting all over again. He was nervous around me and didn't seem to enjoy going on walks with Sherman and me all that much. But since he seemed comfortable with his runmates and caregivers, I didn't worry too much. It looked like he had made a good transition, and I thought he would be OK.

But things took a bad turn for Kaiser in his new situation—he got into a fight with one of his runmates and had to have his injuries treated by the clinic staff. It wasn't long before he was back in the clinic being treated for injuries from another dogfight.

We were learning that Kaiser is the type of dog who gets pretty anxious in a high-energy environment, and that being at Dogtown was too much stimulation for him. When Kaiser gets nervous, he is unsure of what to do or how to behave. The way Kaiser deals with his anxiety is to start fights with other dogs, but a lot of the time he ends up being the one who gets hurt the most. After the second incident, I thought that maybe life in a home with Sherman and me would be a better environment for him, so I decided to foster him and took him home to stay with us.

For Kaiser, it turned out to be the right move. The calmer enviornment plus the increased personal attention seemed to do the trick. He quickly started coming around, warming up to Sherman and me again. He was still anxious, but not nearly as much as when he was at the sanctuary. He still wasn't eating very well, until I found out that he loves cottage cheese. After that it didn't take long to get him up to a healthy weight. Kaiser's new diet and new environment seemed to suit him. His anxiety was lessening, his coat looked healthier, and his bond with Sherman and me was growing stronger.

My dogs go everywhere with me, which means that Kaiser got put into all kinds of new, unfamiliar situations all the time. It helped to have Sherman around, because just like that first day, Kaiser was much braver in these situations as long as he was near Sherman. I brought both dogs to work with me at Dogtown, and Sherman was instrumental in helping Kaiser to feel comfortable in that high energy environment.

We all took vacations together, too. Kaiser's first big road trip was to Washington State. During the entire two weeks, he was practically glued to Sherman's side, but we all had a pretty good time together. Sherman has helped Kaiser get through countless scary situations, and has proved to be the best big brother a dog could have. Things were going so well that I decided to adopt Kaiser.

Being in my home gave him the chance to proceed at his own pace. For the first couple of years, new people made Kaiser nervous, so I didn't push him too hard to make friends with everyone who came to visit. If I had friends over, he would prefer to stay out in the yard until they left. If we went out, it was pretty much impossible to get him inside someone else's house. It used to be that if someone tried to pet him Kaiser would shy away and not come near, but if you ignored him and gave him some space, he would eventually come up behind you to investigate you. When he did become more comfortable (which could take a long time), he would let you touch him. It took some of my friends years to be able to walk up to Kaiser and be able to pet him.

After a few years had passed and Kaiser had grown more confident, though, he started acting more like a normal dog around unfamiliar people Instead of avoiding people outright, he is now comfortable approaching them and checking them out up front. And instead of taking between six months and a few years to be able to touch him, new people can now pet him the same day that they meet him. Kaiser has been with me for almost six years, and I can't imagine what life would be like without him around. People who meet him now find it hard to believe he was ever an unsocialized, scared dog.

As I finish writing, Kaiser is still sleeping on his couch, his legs twitching as he dreams. I can't help but think about how lucky a dog he is to have made it all the way to Best Friends, where we found each other. Every day, thousands of shy, scared dogs just like Kaiser die in shelters around the world. All Kaiser needed was time, patience, and a good dog brother like Sherman to bring him out of his shell. I've never met a dog whom I thought was unadoptable and didn't deserve another chance at life. I just wish there were more homes out there to give the Kaisers of the world the chance for a restful sleep on a couch of their own.

After Knightly lost his first family, his behavior resembled human grief.

Knightly: A Dog in Mourning

Many animals who come streaming into Dogtown are like a throng of mute refugees—tattered and torn, injured in both body and spirit, having survived grim hoarding houses, dogfighting rings, overcrowded shelters, or simple neglect and homelessness. But though their stories may be as heartrending as they are incredible, the details of what actually happened to them are usually almost entirely unknown.

With Knightly, it was different. For one thing, he had a name, unlike all the other nameless arrivals at Dogtown. And as his name implied, he had a noble, almost princely bearing, as if his previous home were not a homeless shelter but the country estate of an archduke. A purebred Weimaraner, Knightly had a smooth body covered by a glossy mouse gray coat and capped by a docked tail. His amber eyes, almost eerily humanlike, reflected his sadness and loss.

His age was also known—he was 13 years old, which is very elderly by dog standards. When he got to Dogtown, he was placed in a run at Old Friends, the area for senior animals, who are older and slower than the others—a sort of canine assisted-living facility.

All of these things were known about Knightly because, unlike many of the other residents of Dogtown, he had spent his entire life with a loving and devoted couple, who treated him as part of the family. It was only quite recently that the trouble had begun. Knightly's owners themselves grew elderly and ill, and when they became incapable of taking care of themselves, nurses and other caregivers regularly began visiting Knightly's home. Like most Weimaraners, Knightly was a loyal and devoted guardian of "his" family, and when these unknown people began invading his house, doing unknown things to his ailing masters, he became extremely protective and extremely confused. He didn't know what was going on, or what to do.

Finally, one day he lashed out and bit one of the caregivers. As a result of this incident, Knightly was taken away from his beloved home and family and placed with a small, independent animal rescue. Distraught, lost, and overcome with anxiety and sadness, he paced and whined, unhappy and uncomfortable. The rescue facility, in turn, sought the help of Dogtown.

And that's how Knightly, the old friend, arrived at Old Friends.

MOONLIGHT ON CHOCOLATE

"When you've had an amazing life, and you are a senior dog, to lose everything seems truly tragic," said Sherry Woodard, who became Knightly's primary trainer when he arrived at Dogtown. "When I looked at him, I understood that he wanted it all back—he wanted back exactly what he had, a life with people he knew, all his familiar toys, a place where he was comfortable. We can't offer his old life back to him. It's gone. But what we do want to offer him—what I really feel I'm obligated to offer him—is a life that is amazing."

Sherry has long, straight golden hair, which she keeps flicking out of her eyes. At Best Friends, where she has worked for 12 years, she is an Animal Behavior Consultant as well as being a nationally certified animal trainer. In fact, she now lives in a house filled with dogs and cats on the sanctuary grounds, just across the street from Old Friends, so that her work and her life are inextricable. She smiled sweetly when she mentioned this.

"I love what I'm doing with my life," she said. "I am living what I believe, and I'm making a difference."

"I sometimes think people might describe me as extreme—that I put myself last, after all these animals. However, I feel very spoiled. I feel very blessed. I feel that I couldn't be living a better life because I'm doing something that I love so much."

When she was a child, she pointed out, "I had incredibly intimate relationships with my pets, like a lot of kids do. But I've held on to my relationships with animals as an adult. I've kept these relationships I had as a child."

And that has become the basis for her life's work.

One of the things that drew Sherry to Knightly—besides the fact that she found him "gorgeous"—was knowing that he had emotional problems.

"People who know me will tell you that I'm attracted to dogs that have been in trouble, dogs with behavioral challenges," Sherry said. "All five of my guys at home are like that. Knightly doesn't really fit that profile—he probably nipped one person in his whole life—but still, he was suffering. He was unhappy. And even though he was a senior dog, even though he did not have much time left on this planet, his life was valuable and worth incredible effort to make sure we offered him a rich life, no matter how short."

According to the American Society for the Prevention of Cruelty to Animals, five out of every ten dogs waiting for adoption in shelters are put to sleep just because no one chooses to adopt them.

Her heart went out to this dignified, elderly gentleman, who seemed so anxious and distraught he just paced in his run, making a high-pitched whining sound back inside his throat.

But even though he seemed as frightened as a child lost in the woods, Knightly came from a long and aristocratic lineage. In the early 1800s, the Grand Duke Karl August of Weimar set out to create a noble-looking, reliable gundog, for hunting big game like deer and bear. He also wanted a dog that would be a loyal household companion, at a time when most hunting dogs were rough customers who lived in packs, in outdoor kennels. The result was the Weimaraner (named after the duke), a "dual purpose" breed who was an obedient hunting dog as well as a sweet, loyal companion. Though the exact ingredients of the breed are not known, Weimaraners are thought to be related to the Great Dane and the red

Knightly's bewilderment at the loss of his family often manifested itself as "chattering," a nervous clicking of his teeth.

schweisshund, a tracking dog related to the bloodhound. They are cousins to the German shorthaired pointer. They have the extraordinarily expressive, slightly morose faces of hounds, but with light amber, gray or blue-gray eyes, which glitter with intelligence. (Weimaraners are so smart they are sometimes called the dog with the human brain.) Pink skin often shows through around their eyes, which seems to accentuate their near humanness. Even their ears are expressive, gray and droopy as a donkey's ears.

But Weimaraners are also powerful, athletic animals sometimes called gray ghosts because they move so silently when hunting. Their coloration is very unusual in dogs (the result of breeding for a recessive gene)—a color that ranges from charcoal gray to mouse gray to silvery gray and was once wonderfully described as moonlight on chocolate.

But the most beloved trait of "weimies," as they're sometimes called—their love and loyalty to a human family—sometimes leads to their most troublesome problems. They're "people dogs," and if they're relegated to an outdoor kennel without much human contact, or left home alone for long periods, they can quickly become neurotic, problematic animals. And if they lose their human family entirely, they can come unraveled, suffering from severe separation anxiety that shows up as panic, anxiety, whining, drooling, or other destructive behavior, which is what had happened to Knightly.

"A dog's attachment to his owner is like a baby animal's attachment to his mother, or a human child's attachment to his mom or dad," writes Temple Grandin, the autistic author of the book *Animals in Translation,* who seems to have an unparalleled insight into animal emotions. When those profound and primitive bonds are broken, animals like Knightly can come as unhinged as a human might in the same situation. People get so attached to their pets they sometimes forget how attached their pets are to them.

Knightly was confused by his new situation and missed his former family immensely. At Old Friends, he continued to pace and whine anxiously, peering out of his enclosure as though he expected his former owners to appear at any moment. He also "chattered," incessantly vibrating or clacking his teeth, like an old man with loose dentures. Chattering, said Sherry, "is fairly typical of old dogs when they're anxious." It's not difficult to empathize with these canine emotions of grief and mourning; they are among the many things humans have in common with our animal companions.

IN NEED OF A HOME

Knightly's vocalizings, Sherry said, were his attempts to express "true emotion—he wasn't faking it. He wasn't calling out because he wanted something, he was calling out because he needed something." What did he need? Sherry felt that Knightly was trying to express that he desperately needed to be in a home, like the one he came from. He was trying to tell everyone that he did not want to live with a big group of other dogs, no matter how clean and spacious the quarters.

Sherry felt that a new home might soothe Knightly's anxiety, but until she fully understood the cause of his anxious behavior—or found a home that was suitable—she didn't think she would truly be able to help him. Even Dogtown would probably not be enough. "A lot of people think of Dogtown as this ideal, perfect place—but it's not that way for all dogs," she explained. "Someone like Knightly, who had this life he loved very much, could be very unhappy in Dogtown."

Sherry knew very well that working with anxiety-ridden animals can be a difficult, frustrating, unrewarding job. One anxious dog who was adopted from Best Friends eight years earlier had experienced "ups and downs" ever since adoption, and Sherry had continued to work with the adoptive family. The dog had "waves of unexplained anxiety" that she and the family had struggled to understand. Sometimes it turned out to be something simple (a fostered kitten), and other times it remained an unexplained mystery.

Ultimately, Sherry wanted to take Knightly into her own home to foster him and get to know him more intimately. But her house was "very special," she said—meaning that it was occupied by five cats and five dogs, all of whom were rescue animals with quirky behavior issues.

As a first step toward taking Knightly home to meet her "motley crew," Sherry decided to try an overnight stay at one of the rental cottages used by visitors, on the grounds of the Best Friends sanctuary. This would be a "trial run" before fostering Knightly in her own home until a permanent home could be found, thereby helping him to become more accustomed to living in a home and thus more adoptable. "I want to get him as close as possible to what we hope to offer him in the future," she said.

When she loaded Knightly into her car for the short trip to the cottage, he bounded over from the passenger side with his big, mournful gray face and gave her a slobbery lick-kiss. "Oh, thanks—this'll be a good date," Sherry said.

Actually, she had already been warned that it probably wouldn't be a good date—a volunteer who'd had a sleepover with Knightly had reported that he had been unable to settle down, whining and following her from room to room all night. Sherry wasn't expecting to get much sleep.

When she got to the cottage, she put Knightly in the house and made two trips out to her truck to get her toiletries and laptop. She deliberately lingered outside for a few minutes with Knightly alone in the house, to see if he might have a full-blown panic attack. (Because he tended to stick to people like glue, never letting them out of his sight, Sherry conjectured that he might believe if he let a caregiver out of his sight, they would be gone forever.) But he seemed to manage this brief separation without trouble.

The ASPCA estimates that owning a medium-size pet dog costs around $695 per year. This includes the dog's food, medical bills, insurance, treats, and license.

When Knightly arrived at the cottage, he willingly walked through the door but still seemed a little on edge and anxious. He paced and whined and chattered. He also expressed his anxiety in another way: spinning around in circles. He didn't zip around like a young dog would, spinning like a pinwheel; instead, the old dog trudged around and around, mournfully, as though plodding along after his own tail.

And Knightly nervously followed Sherry everywhere, shadowing her so closely that she couldn't turn around without bumping into him. Cooking dinner, brushing her teeth, getting ready for bed—the gray ghost was right there, like a Secret Service agent tailing the President. Sherry crawled into one of the two single beds, and Knightly lumbered up after her, awkwardly trying to cuddle. But Sherry needed to keep testing Knightly to see if he would compulsively shadow her. Sherry kept popping out of bed to read, to get something to eat, to check her computer—and every step of the way Knightly followed, dutifully tracking her. Finally, after he seemed to relax enough to fall asleep on the bed, she crept over to the other bed and climbed in. He got up a couple of times in the night to check on his human, then returned to the first bed and fell asleep again.

By the next morning, Knightly seemed to have decided that life was good, he was safe, and nothing bad would ever happen. He actually seemed to be more relaxed in the guest cottage than Sherry was.

"He couldn't have been happier, laying there on that couch," she recalled. "Then he went out and happily rolled around and grunted all over the carpet, with a big dog smile. He was thrilled to be there. He had an appropriate request to go outside. He went to the door. He asked once and then just stood and waited for me to take him outside."

After his sleepover, Sherry said, "I felt much more comfortable with the idea that he could have an amazing future with the right people, in the right home, and that his anxiety would subside."

Still, when she returned Knightly to his quarters at Old Friends later that day, he quickly reverted to his old behavior, anxiously pacing and whining and chattering. "I don't think that we spent enough time together for him to understand that I was coming to get him again and spend more time with him," Sherry said.

Over the following weeks and months, Knightly went to several different people's home for sleepovers as part of a program developed by Sherry years earlier. The purpose was to get the dogs out of their runs at Dogtown, out of their comfort zones, to see how they would behave in a normal home setting full of bizarre and alien phenomena like TVs, dishwashers, beds—and sometimes other animals. The "report cards" filled out after Knightly's sleepovers painted an endearingly mixed picture of him: "So, so sweet," "quiet, a little whiny and clingy, seemed sad, would not relax if I was doing things, he would be standing right next to me," "confused and anxious, will be a great pet with time, patience & training," "wanted to be next to me wherever I was, including snuggling all night long (and even standing next to me as I dried my hair!)." One simply said, "I love him!"

Many of these people mentioned one of Knightly's most touching, and perhaps anxious, bedroom habits—he loved to burrow down under the covers, so that his head was buried all the way down at the foot of the bed.

During his overnight stays with Sherry Woodard, Knightly reveled in her company and a comfortable spot on the couch.

"I think Knightly's messages are pretty clear," Sherry said. "He truly does tell people that he knows what he wants in life. He wants a home, he wants comfortable furniture, he wants people to be near, but overall, I don't think he's needy in an unhealthy way. He seems like he might be a healthy guy, underneath an anxious exterior."

Even so, despite all the sleepovers and kind intentions, Sherry could see that Knightly was not making the sort of progress she felt he could. The "healthy guy" she was convinced lay inside him was still imprisoned by his own bewilderment, grief, and anxiety. And his surroundings at Old Friends didn't seem to be helping him move forward with his life.

THE SEARCH FOR HOME

Finally, Sherry decided to see if it would be possible to temporarily foster Knightly in her own home, as he seemed the most comfortable in domestic surroundings. Knightly's task is as difficult for a dog as it is for a human: Distinguished, elderly, and accustomed to a life of privilege, he would need to give up the past and adapt to a whole new living situation that was more crowded than the one he left behind. At Sherry's house he'd have to share his castle with a gang of other animals who might not consider him Top Dog, as he'd been in his original home.

But before Sherry attempted to bring him home, she needed to know how the old dog would respond to her menagerie of housemates, and vice versa. If Knightly could not find his place in their world, trying to foster him would be unpleasant—perhaps extremely unpleasant—for everybody.

One day Sherry and Elissa Jones, who also works at Best Friends, took four of Sherry's five dogs out to a small, grassy park on the property called Angels Landing. (Sherry had one more dog, a mixed-breed named Trainwreck who stayed at home.) There they would introduce them to Knightly one by one on neutral ground. These introductions would determine if fostering was even a possibility for Knightly; if Sherry's dogs and he did not get along, then he would stay at Old Friends.

Sherry and Elissa carefully orchestrated Knightly's meet-and-greet with each dog. They tried to keep each of the dogs' leashes loose, so they were not adding any discomfort to a potentially tense situation. Sherry had the skills to encourage appropriate greetings between the dogs, even though "lots of dogs lack social skills, and in an uncomfortable situation like this, they may not give appropriate greetings, and then they get stressed and anxious and may start fighting."

Taken together, Sherry's four dogs were a motley crew of rescues. Norton, a long-haired black-and-brindle mutt, was found when he was just a newborn pup, living in a Dumpster on the Navajo reservation; Sherry began bottle-feeding him before his eyes were even open. (Her intention was to foster him, but sometimes "the heart has its reasons the mind knows not of," as Pascal said, and Sherry wound up keeping him. He's now 12 years old.) Chica was a little black dog, mostly Chihuahua, with a graying muzzle, who was rescued from an extremely grim hoarding situation in California. Miles was a fierce little fur ball whose mother, a purebred Chihuahua, abandoned him at birth. Last of all was Shade, a gray, maybe-border-collie-mix with cold golden eyes, who was a former stray found beside the highway.

All of these animals, whom Sherry tended to refer to as "my guys" or "my family," had their odd behavioral issues. But together they were a unit, and if Knightly could not find a way to be accepted into it, the whole idea of fostering him at Sherry's house was a no-go.

First up on the dating game was Miles, whose ferocity was out of all proportion to his size. Sherry called him "my little furry 'gator." Sherry and Elissa gently moved the two dogs together, on leashes, but when they got close enough to make contact, Miles exploded into a fury, or as Sherry described it afterward, "he went into a Jackie Chan doing martial arts all over Knightly's face, with no contact but a lot of noise and a big display." Still, a few minutes later, a second meeting went better. This time Knightly sniffed Miles, tail wagging, which Miles tolerated. Then Knightly walked off imperiously and Miles followed, meekly. "You're doing well, big dog!" Sherry encouraged.

The next dog, Chica, was fearful and tended to snap when frightened, though normally her dog-to-dog interaction was fairly calm. And that's what happened on this day: The little black dog and the big gray dog sniffed each other's nose and backside and seemed content to let one another live.

Next came Norton, who worried Sherry because he was aggressive toward people and other dogs, and had been that way since he was a puppy. She actually took him in because she was afraid to put him out in public. But he was getting older and had mellowed in his senior years. Knightly came up and sniffed Norton, tail wagging, but Norton appeared to ignore him, walking off stiffly. He looked uncomfortable, perhaps feeling threatened. Still, no fireworks. The introduction went well.

Sherry's most devious and unpredictable dog was the gray border collie, Shade. Shade's MO, Sherry explained, was to demonstrate hardly any behavior at all when he was introduced to another dog, "but

Because Knightly grew happier on sleepovers, Dogtown staff concluded that a home environment would be best for him.

then I'll turn my back and hear somebody scream, and that's when he's jumped somebody."

When Knightly approached, Shade went silent. He seemed to be trying to ignore Knightly, refusing to make eye contact. Sherry said this was a bad sign. He lifted his lip just a little, though Sherry didn't think Knightly even noticed.

"He's going to completely ignore him right now, but that doesn't mean he won't offer up something unpleasant when I'm not looking," Sherry said.

"Will you try to be good?" she said to Shade. "Huh? Try to be a good boy?'

But Shade's sneaky golden eyes gave no real clue about what he might be plotting next.

Knightly's reaction to all these introductions was a little difficult to read. He did not seem overtly anxious when he met each animal. But he was not terribly friendly or outgoing, either. He acted mildly curious but reserved.

Ultimately, Sherry decided to take Knightly home to see if fostering him would work. She felt that might be the best way to get him accustomed to living with other animals, and help overcome his anxiety. She was still uneasy about it, though.

"At the end of all our introductions, I still have concerns," she said. "The other dogs are going to need to be supervised. But that won't be a barrier in being able to offer Knightly a place to call home for now, until we find him the ultimate home."

But that place to call home did not turn out to be Sherry's house. Knightly's presence in the house made the other dogs seem stiff, uncomfortable, on edge. Although there was no overt fighting—at least, none that Sherry saw—none of the dogs, including Knightly, seemed able to relax around each other. Oddly enough, Knightly seemed to get along with the five cats just fine; it was the household's canine component that gave him trouble.

The only time Sherry saw Knightly completely unwind was when she closed off the rest of the house, including the other dogs, and allowed the old Weimaraner to lay claim to the entire main part of the house. Suddenly Knightly could exhale completely. Suddenly he felt at home. It was as if he had reclaimed his rightful place as monarch of the realm, and he'd sprawl luxuriously on the sofa, awaiting his next repast.

The other dogs hated this, of course. And Sherry did not feel it was fair to them—after all, they were there first. Unable to find a way to keep everybody happy, after a month or two of trying, Sherry walked Knightly back to his enclosure at Old Friends, just across the street from her house.

She did maintain strong hopes for Knightly's eventual adoption. When she imagined his future adoptive home, Sherry felt it "would be with someone who has a lot of time to offer him. He would love to be with someone who either didn't work, or worked from home, and was there most of the time. He could be a retired dog with a retired family, and that would be wonderful for him."

Dogs have lived with humans for more than 12,000 years in various capacities, including friend and protector.

KNIGHTLY'S NEW HOME

Knightly had a few false starts in his quest for a forever home. A man in Oregon adopted him first, but a change in his personal situation meant he had to return Knightly to Dogtown. Then a woman approached Best Friends about Knightly, but in the end she was unable to adopt him. It was frustrating because everyone knew that Knightly would thrive in a home environment if just given the chance.

But then something wonderful happened. One of the woman's dog-loving friends, whose name was Crystal, had seen the old boy's picture and was interested in adopting him. That's when Knightly finally got what he had been wishing for so dearly.

Today, Knightly has found his home in Windsor, California, in the idyllic Sonoma wine country. He lives there with Crystal, her husband, a winemaker, and two dogs—a Jack Russell-Chihuahua mix named Minnie and a year-old Labrador retriever named Jack. Knightly's house has a big backyard and a neighborhood dog park. He adjusted quickly to these grand new digs, as if he'd finally found his way back to the sort of circumstances he was looking for. Although Knightly didn't spend much time playing with the other dogs—Crystal reported that he was definitely a "people dog," preferring the company of her and her husband—every night Knightly curled up and slept with Jack.

"Knightly is doing very well!" Crystal reported in a message to Best Friends. "I receive lots of compliments on him, he is a very handsome and well-mannered boy. He looks very happy, his eyes are bright and his coat is healthy . . . I am so happy to be able to give him a home, he brings much joy!" Knightly had settled in and finally seemed to relax. It was just as Sherry had predicted: Once Knightly found a home, he would be confident enough to let his true princely self emerge.

Meryl deftly leaps through hoops on the agility course in Dogtown.

Meryl: Lessons in Trust

Watching the big caramel-colored dog Meryl run through the obstacle course called Tara's Run is like going to the circus without having to buy a ticket. Meryl is a gorgeous animal, sleek as a seal, with that combination of grace, power, and agility that only the finest human athletes can approach. In fact, Meryl is a regular agility superstar, effortlessly navigating twisting tunnels, bouncing seesaws, stairs, and tires and deftly weaving through a series of tightly spaced poles at high speed like a daredevil downhill racer threading the gates.

There is clearly a bond of trust and affection between Meryl and her trainer, Ann Allums. Ann guides her through each obstacle, with the gentlest of directions—Ann shows Meryl where to go, and then Meryl enthusiastically attacks each new challenge. Deeply muscled, especially across the chest and shoulders, Meryl is nevertheless lithe as a dancer, nimbly weaving, explosively sprinting, carefully balancing. At the end of the run, she pants happily and lathers Ann all over with wet, sloppy kisses, while Ann rewards her with praise and pats. It's been a good workout for both of them, and Meryl is worn out and content.

To witness the affectionate teamwork between Meryl and Ann makes it difficult—almost impossble—to believe that this dog is under a federal court order to never leave Dogtown. Meryl is one of the dogs rescued by Best Friends in 2007 from a dogfighting ring run by former NFL quarterback Michael Vick. The society and other rescue organizations became involved in the high-profile Vick case when it looked as though all the dogs involved, regardless of history or temperament, would be euthanized. Both the Humane Society of the United States and People for the Ethical Treatment of Animals felt that the pit bulls could not be rehabilitated because they had been trained as fighters, not pets. Best Friends felt differently and took decisive action to save the dogs' lives.

Best Friends and ten other animal welfare groups filed a brief arguing that the dogs could be turned around if only they were given the right environment. They outlined their approach for assessing the dogs and for rehabilitation. It was a thorough plan, one that succeeded in persuading the court to spare the lives of the dogs. Twenty-two of the dogs went to Best Friends (the rest were placed with other rescue groups), but there was a catch. Of these 22 dogs, Meryl, with her history of aggression toward people, and another dog, Lucas (said to be a five-time fight champion), remained under a protective court order: They could be released to Dogtown, but the court deemed them too dangerous to ever leave it.

Best Friends Animal Society (and many others) take the approach that punishment of aggression is the wrong way to address behavioral problems in dogs. According to Best Friends, positive reinforcement is the most productive way to solve dog aggression.

In spite of this prohibition, Best Friends Animal Society accepted Meryl without hesitation because they believed that given the right care, she could turn things around. They were willing to give Meryl a permanent home at Dogtown, where they could get to know her and work with her to give her the best life possible—even if adoption was not in her future.

FEAR OF A BREED

Anyone encountering Meryl for the first time might understand the court's initial fears and the decision to keep her at Dogtown for the rest of her life. She is a formidable-looking animal with a massive, hammer-shaped head, oversize jaws, and heavily muscled forequarters. When behind the chain-link fence of her enclosure, she would sometimes bark relentlessly and snarl when a stranger approached. To anyone unfamiliar with her, the behavior could be frightening. But Meryl's behavior was only one problem. The other obstacle: prejudice against her breed.

The reputation of pit bulls in the past 20 years has taken a hit. Once an adored and trusted companion animal, pit bulls are now believed by many to be terrifying "superpredators" bred for ferocity and aggression and not fit for human society. They have become the dog of choice in illegal fighting operations, which has only done more damage to their reputation. Even the best behaved pit pull, with no history of aggression, must overcome this pervasive bias before being adopted. Because of breed stereotypes, Meryl's barks and growls take on a more frightening quality. But where does this stereotype come from?

In a book called *The Pit Bull Placebo,* vet tech and author Karen Delise argues that every age seems to have a "villain" dog—and at this particular moment in history, pit bulls are it. In the late 1880s, bloodhounds were said to be vicious and bloodthirsty, and no stage production of *Uncle Tom's Cabin* was complete without a pack of baying bloodhounds chasing the escaped slave Eliza. Later it was bulldogs, then German shepherds and Doberman pinschers—especially after World War II when they became associated with Nazis.

Pit bulls (in reality, a vague term that can refer to about 20 different breeds) have been maligned partly because of several frightening myths, Delise argues. One of the most prominent untruths is that pit bulls have a unique jaw structure that allows them to "lock" on to their victims. This is simply untrue, observes Dr. Howard Evans, veterinarian and author of *Anatomy of the Dog,* considered to be the definitive work on canine anatomy: "There is no anatomical structure that could be a locking mechanism in any dog."

It's widely believed that one quality setting pit bulls apart from all other breeds, and even many wild carnivores, is phenomenal biting force, as measured in pounds per square inch, or psi. But one study, which measured (by means of a bite sleeve attached to a computerized instrument) the bite force of a German shepherd, a Doberman pinscher, and an American

Staffordshire terrier (one of the breeds commonly called a pit bull), found that the American Staffordshire had the *least* amount of pressure.

"A TIME BOMB"

It is such misconceptions that cause pit bulls to loom so large in the public consciousness. They have become so frightening to so many, in fact, that an increasing number of localities have enacted "breed-banning" or "breed discriminatory" laws, which make it illegal to own, import, or even possess certain breeds of dog, usually pit bulls or other "bully breeds." Some believe that pit bulls are so dangerous that the entire bloodline should be wiped out; while arguing in favor of a pit bull ban in the city of Denver, Assistant City Attorney Kory Nelson claimed in the late 1990s that "the breed should be terminated as simply being a time bomb waiting to go off."

The wave of fear—some would say hysteria—over pit bulls began to build in the early 1990s, when Denver, Miami, Cincinnati, Kansas City, Toledo, and dozens of other smaller cities enacted bans on them. By 2000, more than 200 cities and counties had enacted breed bans or restrictions. Most of these laws were aimed at pit bulls or dogs with "pit bull characteristics."

Although breed bans are intended to solve the problem of vicious dogs, the issues involved are far more complex than just breeding, when a dog bites a person. In a 2006 *New Yorker* article called "Troublemakers: What Pit Bulls Can Teach Us About Profiling" by Malcolm Gladwell, Randall Lockwood, a senior vice president of the American Society for the Prevention of Cruelty to Animals, summed up the complexity of the problem: "A fatal dog attack is not just a dog bite by a big or aggressive dog. It is usually a perfect storm of bad human-canine interactions—the wrong dog, the wrong background, the wrong history in the hands of the wrong person in the wrong environmental situation."

Lockwood, a leading expert on dog bites, told Gladwell that he never saw any fatal attacks involving pit bulls until the late 1980s, when the breed's popularity began to surge. (Before then, Lockwood said, he saw fatal attacks by every breed of dog except beagles or basset hounds.) It could very

Meryl's life was spared by a 2007 court order, which also deemed that her entire life be lived out at Dogtown.

well be, Gladwell argues, that pit bulls are involved in a fairly large number of attacks primarily because there so many pit bulls—a kind of statistical fluke suggesting that they are more dangerous than they really are.

Which is why, in a nutshell, breed-banning legislation is too generalized an approach to tackling the problem of aggressive dogs. Best Friends' view is that an animal's breed alone should not result in an outright ban or a death sentence. They see dogs as individuals, as products of their training, their personality, their temperament, their level of socialization, in addition to their breeding.

Best Friends' decision to take on a "difficult" dog like Meryl is based on this belief in treating the dog as an individual. Rather than classify Meryl as some sort of demon dog because she is a pit bull, the sanctuary chose to assess Meryl using everything they could learn about her—their professional observations and assessments, her known history from the dogfighting organization, and also the qualities of her breed. There is no denying that Meryl can be a dangerous dog, one who could bite in the wrong situation. But Best Friends believes that introducing Meryl to a better set of circumstances, socializing her to interact safely with people and other dogs, and helping her to learn behaviors that will enrich her life will help Meryl to change. She may not get all the way there, but Best Friends is determined to try. As Ann Allums has described it, "Here at Dogtown is where the dog gets a chance to be who she is, not whom humans fear she *might* be."

BUILDING TRUST

Just because Best Friends is willing to work with aggressive dogs, it doesn't mean that they are naive about the potential risk involved. Ann Allums has worked hard to develop a bond with Meryl, who still has trouble trusting new people and unfamiliar situations. She has devoted countless hours of careful work with her, both for the dog's sake and for her own safety. Ann knows where Meryl has come from and can only guess what she has been through, but the trainer's insight and patience have yielded a strong bond.

The close relationship between the pair is evident whenever Ann approaches Meryl's enclosure. As Ann gets closer, Meryl stops barking, drops down onto all fours, and begins wagging her tail, nosing up to the fence. Her brown eyes look up at Ann beseechingly, as though begging for a treat or perhaps a romp at Tara's Run.

"Hello, Meryl," Ann whispers soothingly. "What's up with you today, big girl?" Ann slips her a chicken treat through the enclosure fence, and Meryl sloppily tries to lick her hand.

Ann lets herself into the run and Meryl, desperate for her attention, playfully bows down toward her front paws, rump up, tail wagging, then dances forward in anticipation of another treat or maybe a back scratch.

Sometimes she jumps up and spins around like a puppy on a spring day. She runs off to fetch a favorite stuffed toy, brings it back, and drops it at Ann's feet. Unlike some pit bulls whose ears have been cropped (some dogs ears are cut for fighting purposes, while others, like Great Danes', may be for cosmetic reasons), Meryl's ears are uncut and floppy, which softens her face, giving it a cuddlier, more endearing quality. Her fur is mostly chestnut brown, with a white blaze on her chest. She is quite young (Ann guesses she is about four years old), and surprisingly, her muzzle and body are almost completely free of scars like some of the other Vick dogs, which indicates she was probably not a fighting dog (Meryl has also shown great affection for other dogs, another sign that she was most likely not used as a fighter).

The Humane Society of the United States reports ten people die in the United States every year from dog bites. It also reports that spayed or neutered dogs may be less likely to bite.

A FRIGHTENING FIRST ENCOUNTER

Dogtown staff first came to know Meryl when they traveled to Virginia in 2007 to assess the dogs rescued from the Michael Vick dogfighting operation, dubbed Bad Newz Kennels (it was named after the neighborhood in Newport News, known as Bad Newz, where Vick grew up). Best Friends sent vet tech Jeff Popowich and dog trainer John Garcia to Virginia to where the dogs were being temporarily housed. They wanted to get to know the animals, assess their needs, and make the transition to the sanctuary in Utah a bit smoother.

Both guys had a special affection for pit bulls, or "pitties" as they call them. John has one at home he calls Spikey Doo. He describes her as "a couch potato" whose "official position in life is making me happy, pretty much." Before he began working with animals, John had no idea what

Ann Allums and Meryl's bond is instrumental to helping Meryl become more comfortable with new people and new situations.

kind of reputation pit bulls had in the wider world, because his experience with the breed had been very positive; the dogs he knew were so eager to please, exuberant, loyal, and sweet. "I just love pitties. I mean, what can you not love about them?"

Even so, Meryl would reveal to John and Jeff a frightening side of her personality. As part of their on-site evaluations, John was walking Meryl on a leash to see how she reacted to several different situations. During the assessment, the pair came upon Jeff, who reached down to pet Meryl on her head. Suddenly, she lunged at him and snapped; Jeff quickly pulled his hand away in time to avoid a bite.

In other circumstances, after an incident like this Meryl might have been branded a dangerous dog and euthanized. But the Dogtown team made a great attempt to understand the incident from Meryl's point of view. Luckily for them, a videographer from Best Friends was on-site and caught the encounter on tape, giving animal behaviorists a way to study what might have triggered Meryl's attack.

The footage showed how Meryl and John were walking together and then approached Jeff, who reached down to greet her. Moments earlier, Meryl had been wagging her tail and checking people out, but when Jeff lowered his hand near her head, her entire body froze. Then Meryl lunged and snapped as Jeff pulled his hand back. It was clear from the tape that Meryl felt threatened by Jeff's seemingly harmless gesture. Though people often reach over and pet a dog on top of the head, many dogs, especially fearful, traumatized ones like Meryl, may be threatened by that movement. In the past, it might have meant that someone was coming to grab their collar.

But something else became obvious when the videotape was shown: Meryl was surrounded by a group of men, all towering above her—not unlike the situation in a typical dogfighting ring. It scared her, and she lashed out. The assessment revealed that one of her triggers was fear of strangers. Despite this incident, Dogtown remained committed to Meryl and her rehabilitation. She made the trip from Virginia to Utah without incident.

PASSING THE BATON

When Meryl first arrived at Dogtown, it became clear that strangers scared her; she showed a lot of fear-based behavior when a new person approached her dog run. Behind the fence of her enclosure, she paced and then threatened with loud, aggressive barks and snarls. The sight of Meryl close to her fence, barking and growling, was, to say the least, unsettling. Meryl's scary behavior, called barrier aggression, may be triggered by two things. First, it may be based on her frustration at being trapped behind a barrier and unable to get out. Second, in addition to being the source of her frustration, the fence may have given her some added bravado, since she knew she could snarl, bark, or lunge against it without actually having to deliver on any of her implied threats. This sort of classic barrier aggression, which is at least partly a function of the fence, often results in dogs being put down in shelters. Potential adopters are intimidated by the barking and overlook the dog.

Another troubling aspect of Meryl's behavior was that she continued to be distrustful of most people even after her arrival at Dogtown. She would warily eye anyone who approached her, and in a few instances, she even snapped at caregivers. Because of Meryl's fear of new people, Dogtown staff went to great lengths to make sure that strangers did not deal with the dog. She wore a red collar, signifying that only staff (and not volunteers) could handle her. Seeing that Meryl had a lot of fear to work through, the Dogtown team decided to take it slow with her, to help build up trust and confidence.

According to the Humane Society, breed bans make it illegal to own a certain breed of dog and are considered discriminatory by animal welfare groups and pit bull guardians, who argue that breed bans are quick fixes that punish responsible guardians by banishing all pit bulls.

One bright spot was that Meryl had learned to trust one person, John Garcia, during the three weeks he had spent working with her before the move to Utah. Frightened of most people, she became attached to this one particular person because he took the time to build up a relationship with her. But John was not going to be Meryl's main trainer at Dogtown; because of her positive attitude, fun methods, and gentle manner, Ann Allums had received the assignment. So from the first day he arrived back in Dogtown, John began working to transfer the trust he had developed with Meryl to Ann, in a process like a relay racer handing off a baton.

Ann welcomed the chance to work with Meryl, as she, too, had a special fondness for pit bulls. "I'm just all the time blown away by how sweet pitties are with people," she says. "And friendly. They just want to please you, they want your attention, they want your petting. They want to just get in your space and sit in your lap and kiss your face, for the most part." But Ann knew that it would take time to bring out the loving side of Meryl, who would have to become familiar and comfortable with her new trainer before she would blossom.

At first, Ann just watched Meryl from a distance, and she noticed two big things. First, she could see how frightened Meryl was of this new place and of all these new people; Meryl still seemed uneasy and uncomfortable, always on guard when strangers were around. The second thing Ann observed was that John was the only one Meryl seemed to welcome or feel comfortable with. Ann noticed Meryl watching John wherever he went. If John started walking away from the dog, she would panic as if she were thinking Where are you going? You're my person! When John approached her run, Meryl would get so excited, Ann remembers. "But John, of course, couldn't be looking after Meryl 24/7. He needed to start passing that trust along to me, right away."

For Ann and Meryl's "first date," John took Meryl out for a walk and invited Ann along. Ann just walked along beside John and Meryl, quietly. She took things slowly, not trying to interact with the dog by petting her. She didn't even look at her, because that could have been threatening to Meryl. The strategy was to slowly get Meryl accustomed to Ann's presence, so that she didn't see Ann as a stranger.

"I am patient," Ann sayid. "I didn't want to rush the situation." (Ann is perhaps a bit too modest. Dogtown vet Dr. Patti Iampietro says of Ann's ability to train dogs: "If they had

opposable thumbs, she could teach them to do the dishes.")

Gradually, Ann's strategy began to pay off as Meryl got used to her being around. After a few days, Ann took the leash from John and walked along with Meryl. She still kept her hands off the dog because she didn't want to scare Meryl and knew that such an action could overwhelm her. A few times, Meryl turned around and gave Ann a look that seemed to say, Who are you? I don't know you. Ann just dropped her eyes and remained calm.

Ann knew well enough that the best way to show a dog like Meryl that she was a nice person was by just leaving her alone and respecting her space. Eventually, in good time, and when Meryl was ready, she became curious about her new friend. She started by cautiously sniffing Ann. A day later, when she sniffed Ann again, her tail started wagging. Ann had shown, through patience and reserve, by waiting until Meryl came to *her,* that she was a friend, not a foe.

As Meryl's confidence has grown, she has become friends with more and more people.

Ann's next step was to start giving Meryl treats through the fence, in the hopes that Meryl would associate good things, like food, with her new trainer. It seemed to work. Before too long Meryl started wagging her tail when she saw Ann coming at a distance. The walks and treats continued until Ann felt that Meryl was ready for the next step: having her enter the dog run.

The first time Ann stepped into the enclosure with Meryl, she was "cautiously realistic," she said. "I took it really slowly." The dog was, after all, a fighting dog who had been neglected and abused, and had a history of lunging at people and being fearful of new situations. Ann was not a new person, but her presence in Meryl's pen was new. It was best to proceed with caution for everyone's sake.

At first Meryl just stood still, looking at Ann head-on. It was a tense moment, but Ann stayed relaxed and loose. Meryl stood her ground while Ann talked and murmured at her, not moving either, patiently waiting for Meryl to approach her. She reached into her pocket and pulled out a treat but did not move any closer. Then it was Meryl's turn, melting out of her rigid stance, dropping her head and moving close enough to take the treat out of Ann's hand. "Good girl, Meryl," praised Ann, rewarding Meryl with a treat and a compliment. Meryl relaxed and before long was following Ann around with eager eyes. Built on a safe history together, they'd begun the next stage of their partnership and were ready to move forward together.

The handoff of trust was complete.

OVERCOMING MERYL'S FEARS

Ann understood that fear was Meryl's biggest obstacle and the cause of some of her scariest behavior. Even though she had formed a strong relationship with Ann, Meryl was still afraid of new people, and she was afraid of her new surroundings. Even when Ann picked up a new object—a silly SpongeBob SquarePants stuffed toy or a grooming brush—Meryl would shy away from the object at first, and only after some coaxing would she lay her ears back and lean in very slowly to investigate. Meryl's body would stay tense, as though ready to bolt at the slightest threat. It was more than a little comical: the pitiless "superpredator" getting ready to flee from a plush, squeaky toy.

"I definitely see Meryl as a dog that we can help," Ann said. "Because everyone she comes in contact with here is good for her. We're gentle with her. We're not threatening to her. And she's a real staff favorite. We're teaching her things that are going to make her a great pet for someone someday, a great companion." Teaching Meryl to overcome fear would be Ann's primary task as a trainer, and Meryl's primary task as a potential companion animal living in the world of humans.

The first step, now that Meryl trusted Ann, would be to develop something called a Life Care Plan for her. The basic idea of a Life Care Plan is simple: to do everything possible to enrich a dog's life for however long he or she stays at Dogtown.

The Life Care Plan documents the activities in the dog's life: how often they get out with volunteers and staff, how often they are walked and taken on outings and car rides, and all the training aspects of their life as well. It documents the dog's progress in learning basic skills, what kinds of rewards they like (treats, toys, etc.), and the speed at which they make progress. If the dog's caregiver moves to some other part of Dogtown, the Life Care Plan, in the form of a fat file, stays with the dog. When a new caregiver takes over training of the dog, he or she can read the chart and pick up where the previous trainer left off. It's an attempt to collect and share information with the large number of people that may be working with any dog, and chart all the dog's training progress, activities, goals, and plans.

Of course, at Michael Vick's Bad Newz Kennels, Meryl and the other fighting dogs also had a "life care plan" of sorts. It's just that the life care plan was one of sadistic aggression training, fear, deprivation, and punishment. Animals were kept chained in kennels, isolated from contact with dogs and humans alike—not really a life at all. They were taken out of their kennels only to breed, to train, or to fight. This plan taught the dogs to be fearful and aggressive and to expect nothing from life. The most remarkable thing about it was that so many of these dogs emerged with their spirits and the ability to trust humans intact.

But when Meryl came to Dogtown, still fearful, jumpy, and mistrustful, her trainers began designing a Life Care Plan for her that would help her to become all she could be. "If a dog is going to be here for their whole life, or just a few weeks, or even a few days, we're going to do what we can to enrich their life," Ann said. "This isn't just a place to warehouse dogs. We're going to make sure that we meet their needs, that we've worked with them, that we give them a quality life. And so the enrichment involves everything from quality food to medical care, physical exercise, mental exercise, rest, socialization, and going to new places. We try to provide that for every dog here."

The Life Care Plan was something Ann was instrumental in organizing at Dogtown. She helped develop a system to organize information about each individual dog, and then this system progressed into a way of providing each dog at the sanctuary with an individualized, customized life plan. "Every afternoon, the caregivers sit down and talk about each dog and chart in their plan how they're progressing. We share with each other, 'Oh, I saw this today in Willie, I saw this in Meryl.' And all of us are interested to hear what's going on with the other dogs, with different people and different situations. It's just a great learning experience, a great sharing experience. We make sure we're on the right track for enriching their lives, and for sharing the good news about a dog's great success on that particular day."

According to the American Temperament Society website, pit bulls consistently score above average on the American Temperament Test for all dog breeds tested.

THE MERYL WORKOUT

When Meryl first came to Dogtown, her muscled body was looking a little flabby. Ann suspected she had been "hanging around the food bowl a little too much." Part of Meryl's Life Care Plan would be to monitor her diet and start her on an exercise program to get her in shape.

Agility training at Tara's Run is one big aspect of the plan. Learning how to navigate the obstacles plus the aerobic activity would do wonders for Meryl's mind and body. But Meryl wasn't a natural her first time on the playground. She was too afraid of the obstacles when she encountered them initially. But Ann knew agility training would be instrumental in helping Meryl overcome her fears—a way to bring out her inner athlete —so she patiently introduced Meryl to each piece of equipment and coached her through each one. Ann knew that as Meryl improved at the exercises, her sense of confidence would grow.

The partnership between dog and trainer would also improve. "Agility training is also great for Meryl because it's a tag-team sport. She's watching me as we go along because I'm directing her to the obstacles that I want her to go to," Ann said. "It's also using her amazing athleticism so we're running over obstacles, jumping through tires, we're going through tunnels and she's running really fast." An athletic dog like Meryl needs an athletic outlet, or else she may grow frustrated and anxious. Ann knew that Meryl would respond well to getting some good exercise because "it gives her something to do."

There is a mental aspect to this physical training as well. The agility exercises teach Meryl obedience to commands, a critical part of her growth. Learning new tricks engages Meryl's mind, giving her new stimulation and interesting things to explore.

But Tara's Run proved to be the first part of the "Meryl Workout." Ann saw that as Meryl's fitness improved, she wanted more exercise. Meryl had so much energy, Ann realized that it would take hours at Tara's Run or a six-mile run to expend all that power. So the next activity to channel Meryl's phenomenal energy and strength was the "mountain scooter." Ann took a simple two-wheeled scooter, the kind kids stand on with one foot and push off with the other, and strapped it on to Meryl's body with a harness. Ann stepped on to the scooter, and then Meryl joyfully galloped along, pulling the scooter and Ann along behind her. After a good long walk, Meryl finally became what Ann called "happy-tired."

After all, she pointed out, Meryl's body was so powerfully packed with muscle, it clearly suggested that, like any human athlete, it cried out to be exercised in order to be healthy and feel good. Because of Meryl's natural athleticism and intelligence, and the evident joy she derives from it all, Ann clearly takes enormous joy in her progress. "I have a permanent smile on my face when I'm not here because Meryl is having so much fun with it. I'm blessed to work with a dog that loves it this much. She's looking at me like, What's the next thing we can do? She's enjoying life. It makes me happy to see her happy."

LOOKING AHEAD

Owing to the federal court order, it's quite possible that Meryl will spend the rest of her life at Dogtown. But a dog could do worse. "We'll always give her a soft bed and a place to stay and great food and attention and love, and companionship," Ann said. "So it's not a home, exactly, but it would be the best home for her if this is where she needs to be. We're committed to her for life."

Meryl has made a great deal of progress in the 18 months since she was admitted to Dogtown, Ann said. When she first came in, she was fearful, stressed, and anxious; it took two or three days to overcome her fear of a new person. Now, after having been introduced to more than 70 staff members and volunteers, it generally takes under a minute before she's wagging her tail and asking for more.

Meryl has a lot to overcome: the fear-based public bias against pit bulls as a breed, and her background as a Vick dog. But Meryl has come really far. She was, in fact, the first of the "Vicktory Dogs" to be moved into a permanent area with a doggie roommate. She spends time with a group of dogs and up to five people in Ann's office. Ann still adores her and spends a lot of time with her—she trusts Ann so implicitly that Ann can do anything to her, grabbing her collar or touching her feet or even (the thing Meryl seems to fear most, for some reason) reaching over the top of her head.

Meryl will probably always need to build a relationship with someone before they can touch her, but her circle of human friends has grown. And though she still wears a red collar, meaning that only staff members can handle her, she'll often greet a new staffer by licking their hand. Sometimes, she still acts like the high-spirited puppy she has tucked inside.

In the end, all the breed-banning talk ignores one critical thing: All dogs are individuals. And Meryl, as an individual and not simply as a breed, has earned the right to be reevaluated by the courts so that she may one day find an owner, a couch, and a home to call her own. Although it's not entirely clear what the courts would need to see to overturn this decision, Ann Allums and Dogtown, as Meryl's biggest champions, are determined to stand in her defense if she has her day in court.

"We aren't quite there yet," said Ann, " but I believe that time will come."

Positive Reinforcement

Ann Allums, Certified Pet Dog Trainer

I have been extremely blessed to have so many special dogs in my life, all of whom taught me something important. There was the dog who taught me not to take it personally; the dog who did bite the hand that feeds him ***while*** feeding him; and the dog who patiently waited while I learned that choke chains were not my training tool of choice. But there is one dog who taught me more than most, the one to whom I owe my dog-training career. That dog is Roy.

Roy was an Akita mix that I adopted when he was one year old. I had just purchased my first house, had a well-paying job, and was ready for a dog. Although I had grown up with dogs and volunteered for the Humane Society during college, I knew nothing about dog training. Roy was a very handsome, energetic, lovable boy—a perfect jogging companion who soon became my best buddy. Despite all the fun we had together, Roy wasn't as easy as the dogs I grew up with. He tended to get carsick. He was a picky eater. He was obsessed with chasing

Frisbees and kids on skateboards. But what concerned me the most was Roy's aggressive behavior toward other dogs and other people.

It started small. I noticed soon after Roy came to live with me that he displayed guarding tendencies by deliberately positioning himself between others and me. (Well, it most likely started with other body language that I didn't know how to read then. I hadn't been around a lot of aggressive dogs at that time in my life, and their behaviors weren't familiar to me.) At the time, it was flattering (and harmless, I thought) to see how much my dog loved me. My friends thought it was sweet how "protective" he was of me. I didn't do very much to correct Roy's displays of "affection," which was my first mistake with him.

After a couple of years, his behavior started becoming more intense and aggressive. He began growling at other dogs and other people, which I ignorantly reacted to by telling him to be quiet. Instead of paying attention to Roy when he was behaving well, I gave him more attention when he misbehaved, which probably reinforced his behavior. Even though I had the best of intentions, it was another misstep in trying to help Roy be a better boy.

Things escalated when he started barking and lunging during our walks. I thought the best solution was to tighten the leash (which I later learned made the problem worse) as we neared other people or dogs. It only seemed to egg on Roy, and his pulling on the leash grew stronger and his barks and growls louder. When friends visited, Roy would often lunge at them if he perceived any movement toward me, even if we were just sitting on the couch together. My attempts to correct Roy's behavior weren't helping things—he only seemed to be getting worse.

I didn't know what to do, so for several years, I just kept him away from other dogs and people. If we could avoid all of those risky situations (like daily walks and visitors to my house), then Roy couldn't hurt anybody. In the process of keeping Roy safely tucked away, I isolated myself as well, and Roy became my only companion. We were safe in my little house together, I thought.

What I didn't realize was that closing us off from the world hurt Roy and it hurt me. We missed the exercise and excitement found on our daily jogs together. We missed the friendship of other people and animals. We were cut off from the exciting and interesting things found in the everyday world because we couldn't take the risk. Roy could hurt someone. I thought it was safer for everyone if we just stayed away.

But even remaining in our house wasn't a foolproof solution. Every so often, Roy would crash through the screen door to chase down a skateboarder, or to attack a loose dog in the yard. I became more and more uncomfortable, realizing that I could not just lock us both away from society forever, and that despite my best efforts to keep him out of trouble, I couldn't control everything. I had a permanent anxiety knot in my stomach, worrying about friends coming over, listening for skateboarders going down the sidewalk, keeping my eyes peeled for other dogs on our outings. I was also full of shame because I had an aggressive dog. I felt like Roy's behavior was all my fault, that I had let him get out of control. Worst of all, I felt helpless to change it.

The turning point came when Roy finally did bite a person. One of the neighbor kids came over to borrow something, and when I turned my back for one second, Roy cornered the boy and bit him on the arm. The image of my dog growling at the kid crying in the corner is seared into my brain. I shamefully took the scared boy, who now had a puncture wound on his arm, back to his mom. For me, I knew I had to take action; feeling ashamed of Roy and his behavior wasn't doing anyone any good, most of all Roy. I wised up and realized I needed to be proactive about his issues. During the next few years, I got an education on different positive training methods that I could use to help Roy overcome his problems. I read as much as I could online, eventually grasping the advantages of reward-based training over punishment-based methods. I visited an animal behaviorist for a consultation. I joined online discussion groups on dog-training topics where other pet parents shared their dog behavior issues. I read the recommended books of those online groups. I attended seminars given by those trainers whose techniques I liked. The more information I got, the more I knew that I had so much to learn!

The concept that resonated in my spirit was that of training to build a relationship with a dog, rather than to dominate and subdue a dog's spirit. I thought I had a loving relationship with my dog—we were best buddies, after all—but I never realized training could be relationship based as well. My previous attempts at training had been to just tell Roy what to do and then expect him to do it. I gave the commands, and he was supposed to obey. When that didn't work, I tried avoidance and shut the two of us away in my house. I saw training as something separate from my relationship with my dog. Training had taken place outside our friendship, but now, based on all I had learned, I knew that training should be a big part of our friendship. Roy and I would share the goal of being able to behave in all kinds of situations. We could resume our daily runs, we could have friends over, and we could both rejoin the outside world. We would be a team.

For the first time, I really saw results with Roy. I marveled at how effective the positive training techniques were! For instance, I began carrying treats at all times and fed Roy the treats in situations where he had normally lunged at other dogs or people. I learned how to pay attention to and understand Roy's body language, so I could read him like a book. When I spotted the signs of early aggression in him, I tried to help him relax. I also trained myself to relax the leash and sing happy songs in those same situations, to reduce tension in both of us. (I started with "Row Row Row Your Boat," and as I got better at relaxing, made up my own hit training songs like "There's a Person, and We Love People!") I rewarded Roy with treats and praise whenever he was successful and tried not to react to behavior I didn't like. It was working. Roy and I were on our way.

To be clear, Roy didn't learn to love other people or dogs, but he did learn to relax. Now he could choose to ignore rather than aggress in more situations every day. I still remember everything about one morning after several weeks of feeding Roy treats as we passed a particular dog on our walk—as we were approaching the dog, I broke out in a happy song and offered Roy treats, but today he declined them. I instantly worried that he was planning to try to attack the other dog. Instead, Roy just charged on—right past the other dog! It was like he was saying, "I got it now. I can ignore the other dog and it's OK."

I realized right then that positive reinforcement was powerful. I saw how it worked to change Roy's emotions. It wasn't about food for him anymore—it was about helping him relax. It was about helping him learn to make good choices. Working with Roy was fun—for him and for me! Using praise, treats, and songs helped me engage Roy and shape his behavior for the better. Roy was a happier dog—he could go for walks without all the drama. Training would remain a valuable part of my relationship with Roy—it made life easier and more enjoyable for both of us.

The training experience with Roy was inspirational to me and set me on a new path in my life—one where I would work with dogs. I always had a passion for dogs, from my earliest memories of growing up with them, but I had not realized that passion extended to training them until I started training Roy out of necessity. As my work with Roy developed, I decided to work toward becoming a professional dog trainer. It wasn't because I was unhappy in my current job as a computer network administrator, nor because dogs can be so forgiving and people so cruel, but because my true passion was helping dogs and helping other people. Seeing how dogs and people can help each other and grow together was more rewarding than just about any other work I'd done.

As a dog trainer, I am learning constantly from the individual dogs I work with, but the lessons I learned from Roy are ingrained in my life as if they are part of me now:

- If you ignore it, it won't go away. Behavior gets worse if treated incorrectly, or not at all.
- Training is lifelong, especially for dogs with aggressive tendencies.
- Praise and rewards bring results. Positive reinforcement methods do work and are ideal for working with dogs with aggression issues.

Because of my experience with Roy, I became a much better trainer. I can empathize now with clients and other owners in similar situations. I remember back to when I first adopted Roy and how bad I was at reading his body language; now I know to stress the importance of understanding how dogs communicate, so that owners can understand what their dogs are "saying."

I also understand the feelings of fear and shame that an aggressive dog can stir up in his or her owners because I experienced that first-hand. I try to help people get past those feelings so they can take action and start helping their dogs. There is nothing like success to make those negative feelings go away, and the sooner an owner begins training, the closer that first big breakthrough will be.

And last of all, Roy taught me that nobody's perfect, least of all me. I loved Roy with all my heart, but I know failed him in many ways. Once I realized that I had the power of positive reinforcement to help me, I did my best to improve our friendship and to give Roy a better life. It is because of him that I am at Dogtown today, where I get to live my passion, and I will always be grateful for the life lessons and changes he effected in me.

Johnny's problems with housetraining made people question his intelligence.

Johnny: Lessons in Learning

"Here, Johnny! C'mon, boy!"

Johnny just stands there.

"Hey, Johnny! C'mon, big fella!"

Johnny looks up slowly, like a crestfallen cow. He doesn't move to the left or the right. He doesn't act as if he's even heard the command. Then—one, two, three, four beats later—he stirs, shuffling toward his trainer, Pat Whitacre, tongue lolling out, with a sweet, dopey grin on his face.

Johnny is a beautiful dog—a purebred golden retriever with such an extravagance of lush blond fur that the groomers at Dogtown line up to brush him. But you don't have to spend much time with Johnny to suspect that he might be a little slow on the uptake. When he first arrived at the sanctuary, the Dog Admissions Report that's filed for each arrival told a tragicomic story:

> *Johnny was raised in a puppy mill . . . owners thought he was deaf due to unresponsiveness... he does not seem to comprehend or respond normally . . . working with him is similar to working with a mentally handicapped child. . . . He is clumsy and when walking by you will walk over you. . . . He loves people and other dogs, is respectful of other dogs' signals, but he doesn't understand when he's done something wrong or something is expected of him . . . he is a sweet, loving, continually happy dog, a "perpetual puppy" . . . he mouths,*

like a puppy, no aggression . . . he lives in his own world, and it's a happy one . . .

Johnny had been bought at a pet store, which meant, unfortunately, that he had undoubtedly been born and raised in a puppy mill (essentially a factory for mass-producing puppies, where breeding adults are often kept in tiny, filthy cages and forced to bear an endless stream of salable offspring).

But after a couple of months, Johnny's frustrated new owners were overwhelmed by Johnny's problems. They complained that Johnny was so dumb he couldn't learn anything. He couldn't be housebroken. Johnny was so slow to respond that they thought he might not only be deaf, but perhaps even partially blind. As a result, he was kept in a kennel, where he would roll in his own feces, making himself both odoriferous and unapproachable. Despite his owners' best intentions, Johnny was not much better off in their home than he'd been at the puppy mill.

The family gave up on Johnny, surrendering him to a golden retriever rescue organization. The breed rescue group, after trying to work with Johnny, also concluded that teaching a slower dog required more time than they could give him. "We've worked with him on 'sit' and it's really hard—he just looks at you," one of the rescue staffers said. The rescue group turned to the first place they thought could help a boy like Johnny: Dogtown. They appealed to Dogtown in the hopes that there would be someone there willing to work with Johnny, who had struck a chord in their hearts: "He is in his own world and it's a happy one! He has a way of looking at you that is so innocent and so loving. We hope that you will consider taking him in."

A SWEET, SLOW BOY

Johnny's file at Dogtown shows that he arrived on November 28, 2007, that he was an 11-month-old neutered male, and that all his shots—parvo, rabies, distemper—were in order. Johnny also showed up with something very few Dogtown arrivals possess: a complete certificate of pedigree, going back four generations. Unfortunately, his file also came with a Dog Admissions Report that basically described him as having flunked out of the elementary school of life. Despite the detailed history, the Dogtown staff would conduct their own assessments to form their own opinions of Johnny's case and what kind of plan was needed to help him learn.

But dog trainer Pat Whitacre warmed up to Johnny right away. He liked him as soon as the big retriever came bounding out of a breed rescue van parked in front of the clinic—a big golden bundle of doggie energy. Pat led Johnny into the lobby and was given the dog's short, unhappy history by one of the rescue staff. "He was this beautiful, happy puppy, who unfortunately had no place to go and a pretty bleak future," Pat said. "His future wasn't too bright unless he could change his behavior enough to be placed in a home somewhere."

Bred a hunting dog, the obedient golden retriever grew to become a popular pet and show dog. Today, the golden retriever remains one of the most popular dogs in the United States.

But despite the history of failure, Pat did not see a hopeless dog in front of him at all. "When people say a dog can't learn, well, it's a challenge to a dog trainer, and I'm a little oppositional by nature, but the statement makes no sense to me," he said. "It's not possible for an animal not to learn unless there is severe cross-circuiting in the brain—unless something's really put together wrong up there."

And Pat Whitacre just didn't believe that was the case.

It's not as though much beloved goldens are dim-witted, as a breed. In a book called *The Intelligence of Dogs,* neuropsychologist Stanley Coren ranks the intelligence of dog breeds, based on the speed with which they learn new commands and several other criteria. Out of 79 breeds, according to Coren's ranking, golden retrievers come in fourth. They're in the top ten group of "brightest breeds" (smartest: border collies, followed by poodles, German shepherds, goldens, and Dobermans; dimmest: bulldogs, basenjis, and Afghan hounds).

But arguments over intelligence aside, one of the things that Johnny really had going for him was that "he's the sweetest dog you'd ever hope

Pat Whitacre was eager to work with a happy-go-lucky dog like Johnny and believed the dog could be taught new tricks.

to find," Pat said. "It's impossible to get him angry with you, and he has no real handling sensibilities, he has no kind of contact reactions from things touching him or brushing him or whatever. He doesn't care who's touching him, he doesn't care if you touch him on his face, or touch him in his ears, or stick your fingers in his mouth, he's not going to snap and bite somebody. He just loves everybody."

WHY CAN'T JOHNNY LEARN?

The fact that Johnny had spent weeks or months in a puppy mill might have had something to do with his house-training problems, Pat felt. Like so many puppies raised in these factory farms for dogs, Johnny wasn't housebroken simply because he didn't need to be. Kept confined in a small cage, he simply defecated wherever he happened to be, and learned to live in his own filth. Also, if he was confined in a small area where he got little if any exercise as a pup, that could have caused a bit of developmental delay in his motor skills, Pat reasoned. "We see that in him even today—he's a little bit goofier and floppier than most year-old dogs would be," Pat said.

Johnny had this harebrained way of stumbling into things so frequently that when he walked through the clinic kitchen, you could sometimes *hear* him walking by the procession of clanks and clatters. He'd do silly, inappropriate things for fun, the sorts of things kids do—like digging to the bottom of the trash can and scattering crumpled paper all over the floor, like someone searching for something.

Johnny, in effect, may well have been in recovery from the crimes that had been committed against him as a helpless puppy.

Pat took Johnny in for an overall physical with Dr. Mike Dix, the head vet and medical director at Dogtown, not long after he arrived.

When Dr. Mike began looking the shaggy dog over, he quickly noted that "he does seem a little bit slow to respond." Johnny would look in one direction, and then Dr. Mike would make a noise and a couple of seconds later Johnny would turn around and look. "There seems to be a disconnect between what happens and how he responds to it," Dr. Mike said. Physically, though, Johnny was a fine specimen—a handsome, healthy dog with a good heart and lungs, able to move freely and without pain. His hearing and eyesight appeared to be normal. Medically, there seemed to be nothing wrong with Johnny at all. He was also, Dr. Mike noticed, "a very nice dog. He's very sweet. He just seems like he's kind of in his own little world."

Dr. Mike described Johnny's attitude as "happy-go-lucky but not aware . . . he's like a rich billionaire who has no clue about life. He's fun. He loves life. He just bounces around being goofy. I saw him playing with another dog one day and he's sitting on top of him wrestling him. He's a goofy dog. He could be a great family dog . . . as long as they don't want too much out of him."

The good news was that Johnny had a clean bill of health. As far as the vets could tell, there wasn't anything physically preventing him from learning. The next step would be on the behavioral side and seeing what could motivate Johnny to learn.

PARTNERS WITH PAT

"Within the first 24 hours of meeting Johnny, I was real impressed with him," Pat said. "This dog gets me very excited. I love this dog."

Pat Whitacre looked at the same animal everyone else had looked at, but he did not see a dumb, hopeless bumbler. He saw a shining star. "He's got so much going for him—so much potential. I can't wait to start working with him. He has so many good features—everything I would look for in a dog to train. He is motivated, he is mobile, he is active, he gives you lots of behavior to play with. And he is absolutely harmless. He is the gentlest dog."

Only a day or two after Pat started working with Johnny, such an obvious rapport had built up between the two of them that other trainers on staff at Dogtown started asking Pat if he was going to take Johnny home and foster him. Pat had already started to think that training Johnny would require being around him a great deal—more than simply visiting him at Dogtown. In fact, Pat felt that full-time fostering would be critical if Johnny was ever to overcome all his bad habits and lack of good ones. On a pragmatic level, Pat realized that fostering Johnny would probably mean not getting much sleep for a number of nights, and if he did get started fostering him on a weekday, he'd be dragging himself to work every day, exhausted. So he decided to wait until the weekend to get started, even though "had I waited one more day, somebody else would've yanked him out and fostered him first, and I wasn't about to let that happen!"

Pat knew that having Johnny in a home environment would be enormously revealing. If Johnny was kept at the sanctuary, Pat would never really know if he'd solved his problems, because Johnny wouldn't exhibit any problems there. In his comfortable, simple quarters at Dogtown, there would be no shoes for Johnny to chew up, no cats to bother, no food on the counter to gobble up—problems that hadn't shown up yet but might in a home._Pat wouldn't even really be able tell if Johnny was housebroken, either, because if he failed to go out through the doggie door and instead defecated indoors on the concrete, it just got cleaned up by volunteers.

In the sanctuary, "There's really no reason for him to change his behavior, especially the behavior that keeps him from getting him into a home," Pat said. "He would basically still be able to continue being Johnny, with very little opportunity for us to help him change those behaviors."

So Pat took the happy, ditzy dog home. Pat led him, still on leash, around his house, where all Johnny's problem behaviors popped up immediately. He tried to chew on anything within reach. He strained at the leash. He tried to eat food on the counter.

At the same time, he seemed oddly oblivious of things dogs usually notice—like Pat's cat. Pat had put his cat safely in a cage so he could gauge Johnny's attitude toward cats, but Johnny didn't even register that there was a cat in the room. Holding out a treat, Pat tried to lead Johnny's nose toward a cat on the windowsill above the sink. But Johnny just playfully grabbed a pot by the handle and started dopily playing with it, oblivious of the cat in the room. "Everything's a toy, isn't it?" Pat said to Johnny, unable to stop laughing at the dog's endearing absentmindedness.

Just because a dog has papers doesn't guarantee he'll have a home! Thirty percent of all animals in shelters are purebreds.

The qualities that had put other people off only made Pat love Johnny more. But although he could see what a great dog this was going to be, he had no illusions about the difficulty of the undertaking. Though Johnny was 11 months old, he had learned almost nothing he needed to know to coexist in a human household, and he had acquired quite a few habits he needed to unlearn. For one thing, Pat discovered, Johnny had no boundaries. He just climbed all over everything, "which is probably more of a problem in a potential home than mine," Pat observed with a laugh, gesturing around his spare bachelor quarters, with the weight lifting bench, dirty dishes in the sink, wood stove, piles of paper, and a general air of cozy, comfortable disorder.

Pat decided that, to expunge Johnny's old, bad behavior and instill new, good behavior, he was going to need to keep a close eye on Johnny around the clock. He decided that the two of them would need to be physically leashed together, at least for the first few days or weeks. He'd even sleep with Johnny's leash around his wrist so he knew when Johnny was coming or going, or needed to get out his doggie door to relieve himself. They were literally inseparable all the time.

The first few days weren't easy. Pat discovered that, contrary to Johnny's first low-key reaction to cats, he actually started to think cats were "kind of cool." If the cats dashed around the house, Johnny would follow, his golden fur a blur behind them, and drag Pat along for the ride. Unfortunately, one of Pat's other dogs tended to get swept up in the excitement and join in the chase, so that cats, dogs, and Pat were all racing after each other through the house. The end of the chase was never as exciting. If Johnny ever actually caught up to a cat, he would just stop on stiff legs and look at it goofily, sniffing, apparently without a clue about what to do next.

Pat also discovered that Johnny had a much less endearing habit, which to most people would be a genuine deal-breaker: He liked to nose around in the kitty box, eating the little "kitty crunchies" he found there. It was yet another bad habit he'd have to unlearn.

THE POWER OF POSITIVE REINFORCEMENT

When Pat brought Johnny home to foster, he thought the biggest problem was going to be the thing that had gotten him thrown out of his first home—his inability to be housebroken. Pat prepared for the worst, fortifying himself with a fresh supply of paper towels and spray cleaner. If Johnny hadn't learned to succeed at this most basic of tasks at the age of 11 months, retraining him could be difficult. Pat knew that, with an un-housebroken dog like Johnny, he'd have to keep an eye on him all the time. "Most dogs will give you some sort of a cue as to what they're about to do, particularly male dogs, who will often have to find an object, sniff on the object, get sideways on the object, and then raise the leg. As a trainer you'll have a lot of warning that he's about to do something; you've got time to hustle him outside before he does it."

But to Pat's amazement and delight, he didn't really need all those paper towels. Johnny was almost—though not quite—housebroken from the second day at his house.

Johnny's goofy antics and sweet nature won him many fans among the Dogtown staff.

When Johnny was about to do his business, Pat picked him up and gently pushed him out his doggie door. After a few tries at this, Johnny would dutifully scramble out the doggie door when the time was right, potty in the outdoor run, and then—get stuck outside. Because the riser on one of the back steps was too high for him, Johnny would clumsily, and unsuccessfully, struggle to get up it. Pat had to fix the step so Johnny could come and go as he pleased.

"I certainly couldn't have him going out and thinking he was trapped out there, and then not wanting to go out any more," Pat said. "Because one of the basic principles of dog training is to make it as easy as possible for him to do what you want him to do."

The primary tool Pat was using to change Johnny's behavior was positive reinforcement, which is at the core of Best Friends' training philosophy. *Negative* reinforcement—that is, punishment for bad behavior rather than reward for good behavior—can also produce behavior changes in dogs. But Pat knew it can also produce unwanted side effects as well. "I don't want Johnny to be afraid of me," he said. "I want him to *enjoy* our training sessions. I don't want him to think, 'I have to shut down and not behave around this guy because if I make him mad, he's going to smack me around or punish me somehow.' Instead, I want to encourage behaviors that I like, which means making them not just fun for me—because I'm tickled that he's doing the right thing—but fun for Johnny too."

Pat would simply ignore Johnny's bad or unwanted behavior: if he jumped up on Pat, or chased the cat, or dug in the kitty litter. Instead Pat only paid attention to the behavior he wanted to encourage, the one he would reward. But for Johnny to learn a whole slew of new behaviors at the same time he unlearned all the old behaviors would require constant, nonstop attention.

"If I work with him 20 percent of a day, he's going to find out the old behaviors are still great 80 percent of the time and the new behaviors may only need to be looked at when he sees me hanging around," Pat explained.

The main thing about a reinforcement that matters is that it should be something that's valued by the dog, Pat said. If he used a piece of chicken jerky and Johnny didn't like it, the whole exercise would be futile. He'd seen dogs who would close their mouth and turn their head when offered a treat—indicating that they wanted to be petted, not fed. So petting and affection is the reinforcer that should be used in that case.

According to Pat, if a dog can do something on his own, he can be taught to do it on cue. If a trainer says "sit" and the dog doesn't move, he can gently *force* the dog's body to sit, force him to obey the command by leading his head with a handheld treat (if that's the reinforcer that works). The trick is finding the dog's trigger—the stimulus that gets a response. For a small boy, it might be a toy truck; for Johnny, it's physical affection, toys, and—Johnny's favorite—treats.

Pat demonstrated this profound, simple principle while Johnny was standing in the middle of the kitchen floor. Pat held out a treat and then waved it around the room. Johnny's nose followed it here, followed it there, and finally followed it into a sitting position because that was the easiest position to reach it from. It was also the place he got to eat the treat.

"He's learning to sit on cue," Pat said. "He's learning, 'My butt on the ground is when the guy lets go of the treat.' This is not a dog that can't learn," Pat said, after he'd fostered Johnny in his home for two weeks. "I can show you a whole myriad of things he's mastered, just like any other dog—I'm kind of embarrassed to say that he already knows more than my other dogs! He literally was sitting on command within 24 hours and then we built on that." Even Johnny's most revolting habit, digging in the kitty litter, yielded to the power of positive reinforcement.

The saying "You cannot teach an old dog new tricks" is false. Shelter animals respond well to good, effective, and humane training techniques. When training your pet, it is important to be consistent, patient, and understanding.

Why had Johnny seemed so slow? Well, Pat says, maybe it was because he *was* slow, and still is—but that didn't mean he was untrainable, or unlovable.

The persistence of the trainers at Dogtown in dealing with a dog like Johnny is genuinely inspirational. Pat Whitacre simply did not give up. He didn't make one of those pragmatic bargains people often make with the universe: "Look, this is a dumb, difficult dog with disagreeable habits, who won't behave and really isn't worth much trouble. Save your energy for a dog that's worth the time."

Instead, Pat simply focused intensely on helping Johnny learn what he needed to know in order to survive in human society—all the behaviors that would make his shining star visible to the world—even if it meant tethering himself to a rambunctious golden retriever all night long. And to what purpose? In order to give Johnny away to a stranger.

Dogtown is, one might say, actually a spiritual community for humans masquerading as a no-kill shelter for animals. It's a kind of monastery where

With Pat Whitacre's help, Johnny proved that was he could learn better manners while staying true to his goofy personality.

the goal is simply giving dogs a happy life, and then giving them away—with no worldly credit for having done so, and only the guarantee of long hours and low pay. For the dog-monks of Dogtown, the payoff is entirely psychic and spiritual. And their good works—in the form of contented, well-behaved dogs—spread out into the world like wave after wave of blessings.

THE SMARTEST DOG ON THE BLOCK

Pat is sitting on the sofa at his house, with Johnny sprawled beside him, his head in Pat's lap. While he talks, Pat absently strokes the resplendent golden fur on Johnny's head and neck. From time to time Johnny heaves a big, contented sigh and looks up, cocking an eye from side to side, as if by means of some cosmic prank he could actually understand every word Pat were saying.

"The biggest measure of how far Johnny has come is that he has come at all—that he is learning," Pat says. "He's learning what he needs to learn to fit into a home, and he's doing it fairly easily.

"Johnny's come such a long way, from having to live in a kennel because his manners were so bad he couldn't be adopted, to living pretty well in my house. He's only been here two weeks, and he's already well on his way to being completely housebroken."

Johnny, the dog people said was untrainable, has come further than anyone believed possible.

"I'm gonna miss him when he gets adopted. He's a very personable dog—he's one of the most popular dogs at Dogtown right now. He loves everybody. You'll never get bored of him watching his antics, whether he's dragging things around or digging through the trash, or whatever he's doing to entertain himself. He's quite comical. He's very well loved."

Not long afterward, when Johnny's story and photo were posted on the Dogtown website, a childless couple from Utah saw what Pat had seen. The couple had lost a beloved nine-year-old golden to bone cancer a year earlier, and the death of their dog had left a hole in their household.

"We are wanting to add a new member of our family—we are wanting to come meet Johnny in particular," the couple wrote, in their adoption application.

When the couple arrived at Dogtown, Johnny greeted them with his sweet, dopey grin, his readiness to play almost any kind of game, his limitless hunger for affection—and a few new tricks, like "sit," "stay," "lie down" and even "roll over."

Rather than being a dumb, slow bungler, unable to learn or even live in a house, Johnny was about to become the new member of a family, and maybe even the smartest dog on the block.

Best Friends rescued more than 4,000 animals after Hurricane Katrina.

Scruffy and Vivian: Staying Afloat After Katrina

In the news photograph, nothing was visible except a sea of dark water and the face of a small dog, terrified and barely afloat, frantically paddling along behind a rescue boat. The photo rocketed around the world because it summed up the plight of thousands of animals stranded, abandoned, drowned, or starving after the city of New Orleans sank into the sea due to the furor of Hurricane Katrina.

By some estimates more than 250,000 animals—dogs, cats, horses, birds, rabbits, fish, ferrets, and every other kind of pet, in addition to zoo animals—were either left homeless or died after the worst natural disaster in U.S. history. In the desperate chaos following the storm, rescue boat captains and helicopter pilots refused to take on animals, reserving space only for humans—which means that the little dog in the picture very likely drowned.

Some pet owners simply fled as the storm bore down on the city, leaving their animals to fend for themselves. Others left food and water for their pets, expecting to be back in a few days; but the days grew to weeks, and many animals starved or broke free and roamed the wrecked city, fighting for survival. Desperate house pets, accustomed to regular meals and a favorite couch, had to revert to living in feral packs on the street, where they were sometimes found covered with chemical burns from polluted water. Many simply did not survive.

Not long after the great storm devastated Louisiana and the Mississippi Gulf Coast, a small, intrepid team from the Best Friends Animal

Society arrived in New Orleans to do what they could to rescue dogs who had been left behind. What they found was appalling.

"We had dogs on houses, we had dogs in attics—we were receiving dogs that had gone through the most horrendous, traumatic experiences I could ever imagine," said dog trainer John Garcia, who was in New Orleans during the nine-month rescue mission. "No wonder half of the dogs we dealt with had some kind of behavioral issue."

By the end of their stay, the Best Friends team had rescued more than 4,000 animals, many of whom were eventually reunited with their owners. Others were placed in shelters or other rescue organizations around the country while efforts to locate their original families continued. About 20 of the most difficult dogs, whose owners could not be found or whose behavior was scary or difficult, were shipped back to Dogtown, in the sunny canyon lands of Utah (at an elevation of 5,000 feet, well above the high-water mark).

Many of these hurricane-damaged dogs showed signs of something similar to post-traumatic stress disorder (PTSD) in humans—psychic scars that can manifest in a variety of different ways. Two of the dogs, Vivian and Scruffy, were badly traumatized by their ordeals. But they appeared to have dealt with their traumas very differently. Vivian displayed what the trainers called "fear-based aggression," but which the average person would probably interpret simply as scary, snarling hostility. Scruffy, by contrast, was pathologically shy and withdrawn, so frightened by everything he hardly had a life at all.

SCRUFFY'S ISLAND OF SAFETY

Scruffy is a well-named dog because he looks, well, scruffy. He was some kind of corn shuck-blond terrier mix, just a little furry fluff bomb who always appeared to be in need of grooming even after he'd just been groomed. His wet black nose jutted out from behind a curtain of his long blond fur, but his pale amber eyes often disappeared behind it.

After being rescued off the streets during the aftermath of Katrina by the Best Friends rescue team, Scruffy was first placed with a rescue organization in Tennessee. Many of the Katrina rescues were placed in foster care or with shelters that made room for them while every effort was made to locate their displaced family. Scruffy's family was never found, and so he stayed in Tennessee.

Life at the animal rescue didn't suit Scruffy and might have worsened his anxiety. Traumatized and uneasy, Scruffy found a safe hiding place underneath a shed in his enclosure where he could go when he felt threatened. When strangers approached, he crept underneath, which made finding an adoptive home very difficult.

The Tennessee shelter eventually closed down two years later, and Scruffy needed to find a new place to stay. Luckily for him and every other animal that came under Best Friends' care during the Katrina rescue, Best Friends had made a lifelong commitment to Scruffy, so he was taken in at Dogtown. Unlike many other traumatized dogs who passed through the sanctuary's doors, Scruffy was not aggressive or particularly dangerous. He was just so frightened and so shy that he could barely breathe. Many dogs breathe hard or pant when they are in a stressful situation, but Scruffy's breath came in heaves and gasps, as though he were hyperventilating. Scruffy also struggled with everyday dog tasks; for instance, he vigorously resisted walking on a leash. In fact, Dog Care Manager Michelle Besmehn discovered how strongly he could resist soon after she started working with him. Scruffy also refused to walk through a doorway into a building.

Hurricane Katrina, the tropical cyclone that hit the southeastern United States in late August 2005, separated many dogs from their families. The storm and its aftermath resulted in the death of more than 1,800 people.

Nothing certain was known about Scruffy's history before the hurricane, and nothing was known about the trauma he'd lived through. One could only speculate. Had he been trapped in a house, perhaps his family home, as the familiar rooms filled with dark water? Or could it have been something as simple as Scruffy having always been an outside dog, who never spent any time indoors? Perhaps the clinic's unfamiliarity scared Scruffy so much that he wanted to avoid it at any cost. If any of these things happened, or anything like them, it would make perfect sense for him to want to avoid buildings. This was not "dysfunctional behavior," it was a reasonable response to his old life in New Orleans, except that now he was safe and sound in Dogtown, and the fear response was no longer needed.

SCRUFFY AND THE SCARY DOOR

One day, with a co-worker, Michelle decided to try to coax Scruffy into the clinic at Dogtown to see how serious a problem this building was for him. Michelle crouched close to an exterior entrance to the building, while a volunteer held Scruffy on a leash nearby. But as the volunteer tried to gently lead Scruffy toward the door, suddenly the little dog stopped short and wouldn't budge. He crouched low, tucked his tail, and avoided eye contact—all signs that signaled panic. As the two people tried to bring him farther along, he started fighting the leash, at first just pulling against it and then finally thrashing his head back and forth, like a furry fish caught on a line.

"This is a big temper tantrum," Michelle observed evenly as Scruffy twisted and turned, his golden hair flying furiously. But she and her co-worker didn't try to force Scruffy inside the building. Forcing him inside could increase his fear and make future training even more difficult. Their aim was to have Scruffy figure out for himself that the door was not a scary thing. They wanted him to enter voluntarily without being dragged against his will.

In another attempt, Michelle and the volunteer tried a different door to the clinic. But Scruffy would have none of it. His reaction was the same—strong and panicked. Trying not to push too hard, Michelle approached him, talking gently. But Scruffy had grown so agitated and upset, he snapped at her. The assessment, they all concluded, was over for the time being. Scruffy would have to learn to enter the door on his own terms.

Bottom line: Scruffy had a long way to go.

"We at least know where he's at, which is that he needs work," Michelle said when it was over. "But I enjoy working with difficult dogs,

Nothing is known about Scruffy's life prior to his rescue, making pinpointing the exact cause of his fears unlikely.

because it challenges me to come up with ideas to help them through whatever is making them uncomfortable."

Still, until Michelle could find some way to help him overcome his traumatic past, Scruffy had little chance of ever being adopted. After all, how could he ever learn to live in a home if he wouldn't come indoors?

Over the following days, Michelle kept slowly, gently working to help Scruffy get off his lonely island of safety and walk through a door. Once again, she knelt in the doorway of the clinic while a volunteer held Scruffy on a firm leash. "C'mon, Scruffy! Go for it! Give it a try, buddy!"

But her encouraging words did nothing to help the traumatized little dog. As before, Scruffy struggled and fought against the leash. In another attempt, Michelle tried offering him chicken treats to lure him inside the building, but they may as well have been rocks. Scruffy's fear of what might happen if he walked through that door was more vivid and powerful than the treats were tempting.

On a different day, Michelle decided to try a new approach—peer pressure. This time, she was going to try to coax him through the doorway using an accomplice—one of Scruffy's dog friends, who could serve as a role model to demonstrate that walking through the door was not the same as walking off the edge of the Earth. This time, when Scruffy was brought on the leash close to the doorway, his dog buddy stood there, also on leash and just inside the door. This time Scruffy seemed torn. His fear and trauma pulled him back; his curiosity and love of company pulled him forward. He whined. He wagged his tail. He whined. Push, pull, push, pull. Forward, back, forward, back.

Finally Scruffy jumped forward onto the small welcome mat just outside the door and stopped there, frightened to go any farther. The two dogs touched noses reassuringly.

"You can do it, Scruffy!" Michelle whispered encouragingly. "You can get inside!"

And then, after days of trying, and multiple failed attempts, almost casually—as if he'd never had any problem with it at all—Scruffy marched through the dark entrance to all his fears.

SCRUFFY AND THE MAGIC CARPETS

Once inside the building, Scruffy faced a new and unexpected challenge: the linoleum floor. Its cool, smooth surface terrified him. So Scruffy found his safe haven on the rug just inside the door. This fear is not uncommon in dogs; the cold, slippery texture of linoleum or floor tiles can make them feel insecure. Though he had made it through the scary doorway, now he was stopped again, paralyzed by a fear of the new surface. He sat huddled on the little square of carpet like a polar bear on a lonely ice floe. Finally Michelle found another small rug and laid it down on the floor next to Scruffy. After several minutes of indecision, Scruffy made his big move, quickly stepping over to the safety of the second carpet.

Cheers went up from everybody who was watching this unfolding story of triumph over fear—a profound human story, as well as an animal story, if ever there was one.

To help Scruffy get used to the floor, three of his dog pals came into the room to reassure him with their presence. The dogs all sniffed each other; and Scruffy seemed to momentarily forget the precariousness of his situation.

Things were going so well that Michelle wanted to remove the rugs to see if Scruffy would stay relaxed without them. When the two rugs were pulled up, it was clear from Scruffy's body language that he was not amused. He crouched low to the floor, almost as if he were trying to vanish into the ground, and he began panting, his tongue hanging all the way out—both signs of fear and stress.

Scruffy seemed to be so panicked that even the other dogs could not distract him, so Michelle ended the session. "If it's not positive and we can't figure out a way to make it positive, it's not helpful," she said.

A poll taken in September 2005 suggests that in the face of a natural disaster, 49 percent of adults would refuse to evacuate from their homes if they could not take their pets with them.

On succeeding days, Michelle began to work with Scruffy's small successes and the "breakthrough" of the magic carpets. To see if she could get Scruffy to explore his new world, she decided to lay down a trail of carpets through the hallway of the clinic, like a series of icebergs in a sea of linoleum. "The idea is, he seems comfortable and confident when he's got the security of rugs, and he seems interested in exploring when he has that," she said. He was like the neurotic Peanuts character Linus with his beloved security blanket, except that in this case, it was a whole Silk Road of security blankets.

In the parking lot outside the clinic, on leash, Scruffy hesitated at the doorway a bit, until Michelle lured him through the dreaded door with chicken treats. Then he began stepping gingerly from one rug to another, as if frightened that the carpets might capsize. When he got to the edge of the last carpet, he peered over the brink as if he were looking into deep, dark water. "One of Scruffy's adoption requirements will be wall-to-wall carpeting," one of the volunteers remarked dryly.

But now, for the first time, Scruffy gradually rose up out of his crouched position and began to wag his tail. He was showing that he was excited to be someplace new. He was discovering the adventure of his own life.

"You can almost see him saying 'OK, I'd like to go over there—could you put a rug there?' " Michelle observed. "So even though he's not confident to do it, you see the desire to do more. It's really fun to see that."

After two months of training, Scruffy had made enormous progress. He could now enter buildings seemingly without fear. And the key to his progress? Rugs. He'd even been able to use them to enter an entirely new place for him—a car. Michelle laid a row of rugs across the parking lot to her car, and then, using cat food, managed to coax him all the way inside the vehicle. Once inside, Scruffy turned around a couple of times and then exited without panic, a very cool customer indeed. "He made the decision to get into the vehicle on his own, which is what we wanted," Michelle said. "The more confident he becomes, the less he'll need to depend on crutches like rugs."

But for now, the carpets across the parking lot, like a trail of ice floes across the ocean, were a way for a little dog traumatized by a hurricane and his former life in New Orleans to come back to life. Today, Scruffy is still living at Dogtown and working to overcome his fear of new situations. Since his appearance on *DogTown,* there have been many applications to adopt the fluffy golden dog, and Best Friends is working to find the best place to roll out the red carpets for Scruffy.

VIVIAN: RED COLLAR RESCUE

John Garcia and others at Best Friends rescued Vivian in the fall of 2005, during the Katrina operation. Vivian's history before the storm remained a complete unknown. Unable to locate her family, Dogtown first placed her in a rescue group in New York. There she acted out her fear by showing hostility and aggression toward everyone, even people trying to help her. She would not stop barking and lunging at the enclosure fence when anyone came near. At one point she cornered a caregiver in a dog run; and when the caregiver reached out, Vivian lunged and snapped at him. After two years, the shelter felt that they couldn't give Vivian the help she needed, so they reached out to Best Friends to see if she could find a place at Dogtown. They hoped that the expert staff there could work with Vivian and make some progress on her aggression.

Unlike Scruffy, whose fear manifested itself in less threatening ways, Vivian was an intimidating presence. She was a powerful, medium-size dog, with heavily muscled chest and shoulders and a squared-off fighting stance. She had a stiff, straight, short-haired tail, and was mostly tan with brushstrokes of white on her chest, throat, the sides of her face, and her feet. But there were a few other mysterious genetic ingredients tossed in for good measure. Her body and shoulders resembled those of a Staffordshire terrier (one of the breeds associated with pit bulls), her large, erect ears looked like a German shepherd's. Others saw in her a bit of Akita, a Japanese breed that in ancient times was used to hunt bears and later used as police dogs. Like most of the other dogs that came to Dogtown, the scent trail of her exact pedigree had been lost in the swamps of the great canine gene pool.

When Vivian first arrived at Dogtown, the staff there got a firsthand look at her threatening behavior. When anyone approached her run, she quickly bounded up to the fence. Her ears back, her body tense, tail erect, Vivian would let out a throaty growl before beginning a torrent of barking punctuated with snarls. Lunging at the fence, she was clearly saying "Stay away from me!" Vivian's behavior indicated that she needed carefully handling. She wore a red collar, signaling that she could be handled only by Dogtown staff.

As with Scruffy, it was impossible to know what terrifying experiences Vivian had lived through on the flooded, putrid, streets of New Orleans. But her behavior seemed to say it all. Vivian's tough act probably served her well in the past. It kept frightening strangers and situations away from her. But now it was serving as a barrier to Vivian, keeping her from helpful people and new situations—the kinds of things that would help put her experiences in Louisiana behind her and move on to a happier, more fulfilling life.

Michelle Besmehn was one of the first trainers to begin working with Vivian when she came to Dogtown. Even someone as experienced as Michelle was a little intimidated by Vivian's show when she first met her in New Orleans: "When I first met Vivian, she was lunging at me through the—through the kennel door, so at first I was a little bit nervous about spending time with her. I wasn't sure what to expect of her, and I had to really challenge myself to be in the right frame of mind to work with her."

Michelle knew she was up to the challenge because she could see that underneath all her bluster, Vivian had a lot of potential to be a great dog. As a young person herself, Michelle had been very shy, so she had a special affinity for animals whose overt hostility was actually masking insecurity and fear and could relate to Vivian's defense mechanisms. "I don't think Vivian is a mean dog," Michelle said. "She just gets insecure in new situations and with new people, and she's trying to tell people that."

A STRATEGY FOR VIVIAN

Michelle realized that there was a window of opportunity for training Vivian because she had just arrived in a new, scary situation where she didn't know anybody and felt doubly insecure. Michelle intended to take advantage of this, using Vivian's fear to help build a relationship with her.

To start, Michelle began by talking gently and feeding Vivian treats through the fence. Gradually, she taught her to sit in response to her command. Michelle had to be assertive and firm while staying positive, ignoring negative behavior, like Vivian's barking and lunging, and only rewarding the behavior she wanted.

As Michelle began breaking through Vivian's wall of hostility, she discovered that Vivian's personality could sometimes be regal and standoffish, and then at other times goofy and puppyish. Vivian had to be completely at ease to reveal these sides of her personality, so most casual observers didn't get to see them. To most new people, she remained aggressive and hostile—but it was just a front.

Growling and barking, Michelle said, are two forms of communication—two ways for dogs to say they are uncomfortable with a situation. They are also warnings, a way for a dog to say, "Get away. I'm uncomfortable with you being that close, and if you don't get back, I'll bite." Michelle's goal was to make Vivian feel comfortable with new people, and new situations, so that she wouldn't need to uses those forms of communication.

As the relationship between Michelle and Vivian began to deepen, Michelle was able to slowly and carefully expose the dog to new things. The treats allowed Michelle to earn enough trust to safely enter Vivian's run (although there still were occasions when Vivian would stiffen and seem unsure when Michelle entered her pen, perhaps as a prelude to biting; Michelle would have to force herself to stay calm and wait for Vivian to relax). Eventually Michelle could slip a leash over Vivian's head and take her for walks. Vivian began to see Michelle as the person who gave her good things, including treats and walks. Michelle discovered that Vivian loved toys, and she'd sometimes pick up a toy and throw it toward her, to help the big dog relax—as if to call out Vivian's inner puppy. Only then could she get close to her and get her to mind.

According to a news report, events such as Hurricane Katrina might affect dogs as well as humans, resulting in behavioral symptoms that may be similar to post-traumatic stress in humans. The canine symptoms include barking for no apparent reason, hiding, forgetting potty training, losing weight, and even behaving aggressively.

"I think Vivian bonded with me because I made the effort," Michelle said, "and because I called her bluff."

VIVIAN'S "FENCE FACE"

Michelle had been working with Vivian for three weeks, and making steady progress, when she decided to introduce her to trainer John Garcia. Vivian had shown a marked preference for female caretakers over male, and Michelle knew that getting her comfortable with both men and women would be important in her rehabilitation. Seeing how she reacted to John and if she was capable to relaxing around him would be the first step.

It was a snowy day at Dogtown when John and Michelle approached Vivian's run. Vivian began lunging ferociously at the fence, a behavior known to the trainers as barrier aggression, meaning that the very presence of the fence tended to make the dog more hostile and threatening. Vivian would actually back up and then charge the fence line, barking loudly.

"She's got a really good 'fence face,' " John said calmly. "She gets the reaction she wants, which is for people to leave her alone and not go into her run."

Most people would be completely intimidated by this display, if not terrified. But John and Michelle simply decided to listen to what Vivian was saying. By reading her body language and listening to her barks, both John and Michelle felt that Vivian was saying, as clearly as she could, that if John had entered the run at that point, she would probably have bitten him.

And John was listening, because although Vivian had no bite history, John had been bitten badly—twice. Once he was inside a run with two dogs who were squabbling when somebody asked him a question and he was distracted for a moment; in a flash one of the dogs bit him so severely on his left calf that it took 60 stitches to close the wound. He was on crutches for a month.

The other time, he was bit by a dog he'd been working with for seven months and trusted completely. He'd brought the dog into his own home, taken him for long walks, and rolled around on the ground with him. He was in the process of putting the dog into a crate when suddenly "he just grabbed my right hand and sucked me into the cage." The dog held on for three minutes. John spent two weeks in a hospital, but afterward his hand was still sore. It turned out he had a bone infection, requiring still more surgery. The hand still bothers him every day, he said. These two instances remind John daily how important it is to be vigilant with dogs, even ones he knows very well.

Because Vivian can be difficult to manage, she wears a red collar, signifyng that she may only be handled by Dogtown staff.

"I would be a liar if I said I'm never afraid of dogs," he said. "I mean, fear keeps us alive. For the most part I just try to use common sense so I don't get myself hurt."

John remembered how powerful Vivian had been when he had to hold her down for a blood draw during her medical intake. She'd been muzzled, but she fought mightily. "She's probably one of the strongest dogs I've ever held in my life," he said. John knew it would be unwise to test her by entering her run when she was telegraphing all the signs of a potential bite.

BECOMING PART OF VIVIAN'S LANDSCAPE

For this meeting, despite all her threats and bombast, John decided to react to Vivian by not reacting. He just stood there, a few feet back from the fence, as Michelle tried to soothe the dog.

"I was just part of the landscape," John said. "I wasn't a person, I wasn't a dog, I wasn't anything. I was neutral. I was zero. It's much easier to introduce yourself to a dog that way.

"When I first started working with dogs I didn't have that ability to control my emotions like that," John said. "When a dog threatened me I basically got scared. But that will magnify the dog's behavior because it makes them scared." But by keeping his emotions in check, over several meetings Vivian became more comfortable with John.

John continued to build a relationship with the skittish dog. Like Michelle before him, John used treats passed through the fence to nakedly bribe Vivian. That way, she came to associate him with something positive. Whenever she did something positive—like approaching the fence calmly, coming when called, or sitting on command—Vivian's ear would push forward, wrinkling her forehead, as she attentively looked to John for a tasty treat. Rewards distracted Vivian from trying to scare John and focused her on trying to please him instead. It didn't hurt that John would also sweet-talk the big girl shamelessly. "There ya go, you're so smart, not to mention beautiful—you're my type of woman, yeah . . ."

Eventually, John got to know Vivian well enough that he and Michelle decided he should try to take her for a walk. Because of Vivian's history of aggression, Michelle was one of only two people who had ever taken her for a walk at Dogtown. Now John wanted to become the third.

It was another snowy day at Dogtown when John and Michelle entered Vivian's run. Though John was a skilled dog handler, he kept a wary eye on Vivian because he had seen how "without any warning, she'll sometimes snap and try to bite." Michelle leashed up Vivian and led her out of the run down a snowy trail, with John walking very casually alongside, a couple of steps outside biting range. Almost imperceptibly, Vivian's stance began to loosen. She seemed to grow more relaxed, and more interested in reading her "morning newspaper"—all the markings from other dogs along the trail.

In a soft voice, John asked Michelle to hand over the leash, and she did. The handoff had been barely noticeable, but it was critical—John called it "the make or break moment." Vivian glanced back and saw this, made note of the fact that she was now being controlled by a different human, and moved on.

"How's it going, baby girl?" John said soothingly. Then, to Michelle, he said, "She's looking right at me, so she knows I'm walking her. She's gravitating toward you, but at least she's acknowledging me as safe and secure."

After that, Vivian just seemed to be ignoring John, which is exactly what he wanted.

The Federal Emergency Management Agency (FEMA) provides tips and strategies for caring and preparing pets and other animals for emergency situations. You can visit it on the Web at: http://www.fema.gov/individual/animals

NEW FRIENDS FOR VIVIAN

Later on that same afternoon, Michelle and John decided to attempt to move Vivian into a run with two other dogs who also had traumatic histories—Joe, another Katrina survivor (who had been returned from an adoptive home for biting), and Pedro, who had been rescued from the war zone in Lebanon. Moving a dog like Vivian into a group situation could potentially help her learn to live with other animals. But the trainers were also aware that putting dogs together like this could be explosive, like throwing a match into a fireworks factory. In fact, John said, "Dog intros are probably the most dangerous thing we do here."

Still, John and Michelle knew that both Pedro and Joe had good "dog skills," meaning that they mixed easily with new dogs, and that Vivian also liked other dogs. (Her hostility was directly primarily at humans.)

A good scratch cements the strong friendship between Vivian and Dogtown manager Michelle Besmehn.

First, Michelle went into Vivian's run and leashed her up. Vivian knew Michelle well enough by now to tolerate this amiably. Next, John leashed up Joe, a black Lab, and walked him into the run with Vivian. The two dogs sniffed each other curiously. Then Joe began to romp around Vivian, in a puppylike way, darting in and out. Vivian stood her ground, regarding him with something like regal amusement.

"Boy, she's got that 'tough girl' attitude, doesn't she?" John said.

Then it was Vivian's turn to dart around playfully, seeming to accidentally bump into Joe. Vivian was clearly the heavier, more powerfully built of the two dogs, and this body-bumping play-display may quickly have established a pecking order. Whatever was going on, it seemed to have satisfied both dogs, and they both settled down surprisingly quickly.

Next Pedro was brought in on leash, and a similar getting-to-know-you routine took place. The three dogs seemed to quickly work out a friendly social arrangement. It appeared that Vivian was becoming a part of the pack.

This placid scene seemed a world away from the first day John saw Vivian savagely lunging against the fence. "The first time I saw her, she had the most negative reaction towards me imaginable," he recalled. "She was really nasty. And now, she sees me and she gets all wiggly. She really is happy and enthusiastic to have me take her out of her run. Once you get to know her, which isn't that difficult, she's a lover. I mean, she's just adorable."

But Vivian isn't out of the woods yet. She has made great progress in overcoming her fears, but there is still much to accomplish. She must make new connections with new people and fully overcome her barrier aggression before adoption can be a reality for her. Vivian continues to live at Dogtown as a red collar dog. She and her trainers, including Michelle and now John, will continue to work with her on her barrier aggression and other threatening behaviors.

John said, "I have high hopes for Vivian because she is such a cute, loving dog, but no matter what, we're not gonna say 'Oh, she's fixed. We'll put her into a home with kids' before she's ready. We're not gonna set her up to fail. We're gonna set her up for success. Even if Vivian is never placed in a suitable home, we'll take care of her here at Dogtown forever. The rest of her life is gonna be a good life, no matter what."

The fear and chaos of Katrina, which spread like dark water over the lives of Vivian and Scruffy, may have stained their spirits forever, but the storm also brought them to Dogtown, where these gallant creatures may learn to live. There, they can be "set up for success" despite all the psychological wounds they bear. Their pasts have given them much to overcome, but the dedicated team at Dogtown will continue to work toward their ultimate goal: a life without fear.

Infamous Spikey Doo

John Garcia, Dogtown Manager

Over the years there have been many, many dogs who have changed my life and in some cases even saved it. But the one who will always be closest to my heart is my girl, Spikey Doo. I met her when I was 16 and visiting friends in southern California. Back then, her name was just plain Spike. She was a year-old American pit bull terrier (I call them pitties for short) and had a great life. She had lots of room to run around in her California backyard. There were lots of little kids around to play with her, too. Her rambunctious, fun-loving personality jumped out at me, and I instantly saw what a great dog she was.

But Spike's situation was changing. Her family had to move to Texas and couldn't bring her along because of housing restrictions in their new place. They had exhausted all their options in trying to find different living arrangements and were really concerned about what was going to happen to their dog. Before the move, they were looking high and low for a forever home for Spike, but time was running out.

As soon as I met Spike, I thought she would fit in perfectly with my family back in Utah. Because my mom and I lived out in the middle of nowhere, there was a lot of room for Spike to exercise. We also had a lot of companion animals for her to play with. At the time I had a dog, a chow-timber wolf mix named Sprocket, two cats, ducks, chickens, a rooster, and a crow. Plus, in my mind I could just imagine how much fun Spike would be as a hiking buddy. I had always had energetic dogs like her in my life, so I told the family, "If you can't find a place for her to go, let me know. She can come live with me in Utah."

Of course, after I made the offer, in my mind it was set in stone: Spike was now my dog. But there was another person I needed to consult: my mom. I hadn't yet let her know that another dog was on her way to our home, so I decided to do the right thing and tell her. Mom wasn't very surprised; I would routinely show up with abandoned animals to care for. This time, I could tell she was hesitant with me bringing home a strange dog, but being the best mom in the world, she told me she would help me transport Spike to Utah. My mom told us to rent a car, drive Spike out to a halfway point near Las Vegas, and that she would meet us there and drive Spike and me back to our home.

Spike's family and I rented a swanky Lincoln Town Car with all the bells and whistles (the first one I'd seen) for Spike's road trip to Vegas. Spike's family decided to come with us so they could say goodbye to her. Even though they knew Spike would have a great home, they were still very sad. It was an amazingly difficult decision for them, and they felt as though they were truly losing a part of themselves. I kept assuring them that I would take very good care of their baby and they were welcome to visit her anytime. There were a lot of hugs and tears when they said goodbye. And then they got in their car and drove back to California, and we got into ours and drove Spikey Doo to her new forever home. Once we arrived in beautiful southern Utah/Northern Arizona (yes, I'm actually an Arizonan), my adventure with Spikey Doo really began.

Being 16 and not really knowing much other than what I had experienced with all my other dogs growing up, I was under the impression I would just let Spike out of the car and she would fit right in with all the other animals at our house. We were one big happy family, and I thought Spike would just blend in with everyone. Boy, was I wrong.

Spike's introduction to our pets didn't go very smoothly, to say the least. Spike didn't know other dogs and cats—much less ducks, chickens, roosters, and crows—so she didn't know quite what to do around them. Looking back, all I can say is that Spikey Doo must have thought I'd brought her to her first buffet. The first thing she did was try to pick a fight with Sprocket, and then proceeded to chase anything that would run away from her. It was chaos—every animal was sprinting away from her. I ran after Spikey Doo and finally caught her before anyone got hurt. As I stood there catching my breath, I started to think that I might have made a very bad decision.

My mom looked overwhelmed as she watched. She knew how important it was to me to save Spike's life, and she could tell from the look on my face that she wasn't the only one overwhelmed by the situation. But, great mom that she is, she took one look at me and told me not to give up. If we kept trying, we would find a way to help Spikey Doo adjust.

This was one of my first experiences with truly negative behavior in an animal. All of my other dogs (including pitties) were relatively issue free because they had been exposed to lots of them from a very young age. Growing up, I always felt like I could handle anything that came my way when it came to dogs, but this was the first time I came up against big socialization issues like these. It was all new to me.

The goal was to get Spikey Doo used to our other pets, and I was determined to help her do it. But I wasn't quite sure how. Spikey Doo clearly hadn't been around a lot of other animals in her California home. Because she lacked socialization, she had no experience to draw on when she met these other creatures. When I was pondering how I could help this new spunky girl while not jeopardizing any of my other animals, it hit me—I simply can't give up.

One of the first things we tried actually turned out to be the best: constant supervision. I stayed with Spikey Doo every minute—everywhere I went, she went, too. It was a lot of work because I had to make sure she didn't put any of our pets in danger, but it was also a lot of fun—we really got to know each other. She was a playful, smart, spunky girl, who learned quickly but who still needed a lot of guidance when interacting with other animals. The first few weeks were rough, but I stayed patient and gave her lots of time to figure things out. To this day, I think the best thing I did was giving it time and letting her learn that all the other animals were part of our family and, now, so was she. As the hard work started to pay off. Spikey Doo eventually understood what I was trying to convey to her and started to see the other animals as friends, not lunch.

That one-and-a-half-year-old pup has grown into a gray-muzzled, little old lady. Spikey Doo has come a long way since those first sketchy days, and, believe it or not, today she is a role model for other dogs with socialization issues like the ones she had. She routinely helps new foster dogs coming into our home by "showing them the ropes" and being a positive role model. New dogs look to her example to see how to act.

Life without Spikey Doo is unimaginable to me. Helping her become the wonderful girl she is today was one of my first true challenges, but all it really took was patience, love, and understanding. Spikey Doo gave me several crucial lessons about dogs. Through Spikey's first meeting with my pets, I saw how dogs truly are individuals—just because all my other dogs liked other animals didn't mean that every dog would behave the same way. I learned that the most powerful tools in helping create change are patience and understanding. Once I took the time to see the world through Spikey Doo's eyes, it helped me to figure out the best way to help her. And my experience with her put me on the path to Best Friends and to a career dedicated to saving dogs just like her. I owe her a great deal and am thankful for every day I get to spend with the infamous Spikey Doo.

Wiggle's cute face and a winning personality are impossible to resist.

Wiggles: The Cutest Ugly Dog

Wiggles, a little fawn-colored bulldog mix, was a year-old pup when an animal control officer picked him up wandering the streets of California in 2006. He might as well have been arrested for public drunkenness, because the most noticeable thing about Wiggles was that he couldn't seem to walk in a straight line. With his bottom jaw jutting out and his black-spotted tongue lolling out of his mouth, he looked like a rum-sodden sailor, always listing to starboard, crashing into things, and sometimes just plain falling over. Animal control was charmed by the young dog's silly personality; they kept Wiggles with them for a few months but couldn't find him a home. The shelter in California knew that Wiggles might have some serious health issues that they couldn't handle, so they called up Best Friends, which decided to take on this dizzy stray dog.

When Wiggles arrived at Dogtown for diagnosis and treatment, his quirky looks and wobbly walk very quickly began winning over the volunteers and staff. His listing gait and his funny face were only part of the package. Wiggles was a sweet, friendly boy, who enjoyed people and attention. Nobody seemed quite able to resist him.

It was Wiggles' face that was the killer. The whole thing looked as though it were halfway smashed in, in that classic bulldog way, as if he'd collided with a wall at high speed. But he had an immense protruding underbite displaying a rack of canines that looked as though they were intended to stop fights simply by looking so scary. Yet at the same time, there were these sweet, expressive brown eyes, without a hint of threat in them, and those erect, forward-pointing ears, one up, one flopping down.

The contrast between those amiable eyes and the predatory underbite was so comical and charming that you couldn't help smiling when seeing Wiggles wobble around his enclosure at Dogtown. The ferocity was not quite believable, being in such close proximity to the adorableness. Maybe that's why some people think bulldogs are the ugliest dogs, others think they are the cutest dogs, and still others think the combination makes for one of the most endearing faces of all dogs.

"My first impression of Wiggles was that he's just very cute in a very ugly sort of way," said Dr. Mike Dix, head vet and medical director of the clinic at Dogtown. Dr. Mike, a serious-minded medical man, was primarily intent on making a medical assessment of this odd little animal when he was brought into the clinic. Still, he couldn't help it: His heart went out to Wiggles.

"He is easily the ugliest, cutest dog I've seen in a long while," he said. "He's got charm. He's goofy. He doesn't just walk goofy, his personality's a little goofy. He's just very . . . endearing."

"I LOVE IT 100 PERCENT OF THE TIME"

When Mike Dix was growing up, there was always a family dog, starting with a German shepherd, followed by a black Lab, who became Mike's very own dog, and with whom he had a very close relationship. He always seemed to have close relationships with animals. When he went to visit his grandparents, he'd spend more time playing with their dogs than talking to them.

When he got a little older, he decided he was going to be a football or basketball player when he grew up. One big problem: "I had no talent." So by high school, Mike decided to turn his love for animals and love of science into a career as a veterinarian.

It was Dr. Mike's wife, Elissa Jones, who was instrumental in bringing him to Dogtown. He had a busy veterinary practice in Portland, Oregon,

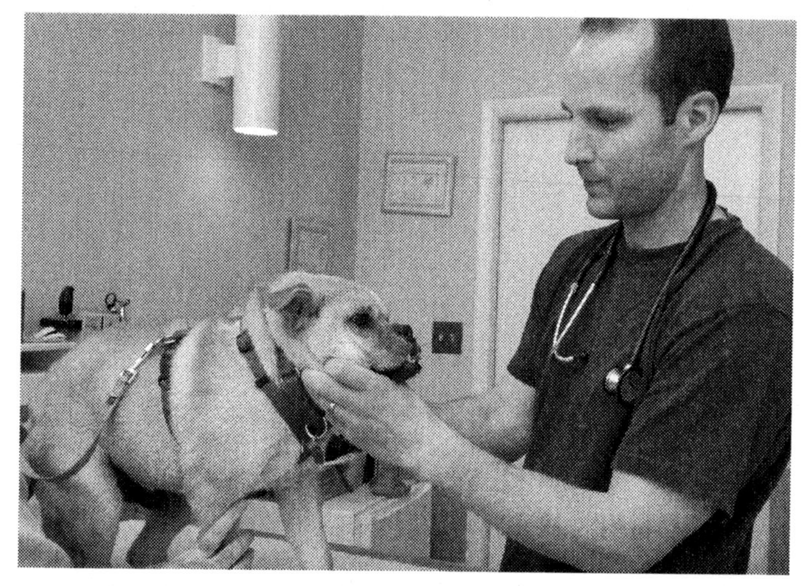

Dr. Mike Dix, the head vet at Best Friends, examines Wiggles to diagnose the possible cause of his unstable walk.

until a few years ago when Elissa, who was a devoted animal lover and a longtime member of Best Friends, started trying to persuade him to come to the sanctuary in Utah for a volunteer vacation. Elissa had come to Best Friends for several vacations and had adopted a dog there, and she loved the place.

"I'd say, 'Sure, I'll visit, but I work with animals all day long, so I'm not going down there on vacation and do the same thing,' " Dr. Mike told her. He also wasn't a big fan of the desert: hot, flat, and boring. Or so he thought. He kept resisting.

"But you know, I loved Elissa, so finally I came. And when we got here, it was actually a very gorgeous place. Everyone at Best Friends was great. I liked the work they were doing and they just happened to be hiring a vet while I was here, so they asked me to spend some time in the clinic and I did." When they offered him the job, he took it.

Now that he is medical director of the busy clinic at Dogtown, he says, "I love it 100 percent of the time. I don't stop smiling all day."

"One of the most fulfilling things about being a vet is that I am just working for the animals, which is a wonderful feeling. That's what I got into this business for. But one of the most frustrating and interesting challenges of veterinary medicine is the fact that the animals can't tell you what's wrong with them."

Despite his touchingly expressive countenance, Wiggles couldn't tell Dr. Mike what was wrong, whether he felt better or worse, or what he wanted. Dr. Mike had to figure it out, and decide the best thing to do for him.

HUNTING FOR CLUES

Wiggles' exact breed was difficult to determine since he had been picked up as a stray and there was no preexisting medical history for him. He appeared to have some bulldog in his background, but just how much could be not be ascertained with any certainty. Like most of the animals who came into Dogtown, Wiggles was more likely an unknown jambalaya of breeds and half-breeds all stirred up in a pot. Sure, he was probably *mostly* bulldog, but there were enough other ingredients in the secret sauce that Dr. Mike referred to him as "bulldoggish."

On the other hand, one thing Dr. Mike noted during his initial medical evaluation was that in the California town where Wiggles was found, there had been a bulldog breeder. Some of Wiggles' more "distinctive" traits resembled problems caused by inbreeding. It's possible that Wiggles was a purebred dog gone awry—an attempt to achieve bulldog perfection that failed. Dog breeders sometimes use inbreeding to guarantee desirable traits—for instance, by breeding a male with brindle coloration (considered desirable) with a female of the same color. But this technique limits the gene pool and can lead to genetic mutations, with symptoms like the ones Wiggles displayed. His topsy-turvy drunkenness and underbite might have been a problem caused by bad breeding.

During his initial exam, Dr Mike had to determine if Wiggles suffered from a lifelong, possibly progressive illness caused by this breeding snafu, or if his condition was manageable. Wiggles sat quietly on the examination table and patiently submitted to a battery of diagnostic tests that would determine if his problems were neurological or if they were caused by an illness. Though he could not communicate where he hurt or what he needed, Wiggles sat patiently on the exam table, his eyes expectant and trusting. His body may have been bulldoggish, but his heart was about as ferocious as a lollipop.

First, Dr. Mike took Wiggles' front paw and turned it over, so that the "palm" was facing up and the back was on the table. Normally, a dog would quickly right his paws. But Wiggles just let them sit there, as if unaware of them. His reaction indicated that there was a disconnect between the nerves in Wiggles' foot and his brain. The location of the problem—in the brain, the nerves, or the pathway to the brain—was unknown.

Dr. Mike suspected there might be something going awry in Wiggles' cerebellum, the part of the brain that controls balance, which also plays a vital role in basic motor function. So he performed another simple diagnostic test, pretending to suddenly poke a finger in the dog's eye. Each time, Wiggles blinked. This normal blink response meant that his cerebellum was normal, which indicated that the problem lay somewhere else.

At the Dogtown medical clinic a staff of 20 cares for the animals who temporarily live in its 25 cages (the equivalent of hospital beds). There is also a dental suite and an x-ray lab at the clinic.

Next, he x-rayed Wiggles' spine to see if it was malformed. Surgery could fix a curved spinal column if that was the source of Wiggles' problems. But an examination of the x-ray films on the viewing box ruled that out. "His spine looks pretty good," Dr. Mike said. "That doesn't mean it's not a spinal issue, but there's nothing obvious. So with a severe brain or spinal injury ruled out, it's likely that Wiggles was born with some kind of genetic defect that affects his neurological system."

"Now if we ran every test possible—CT scan, MRI—and decided whether it was a brain problem or a spinal cord problem, you might find something surgically that you can repair. But is that really going to be in his best interest? I tend to be a minimalist for a lot of these cases,

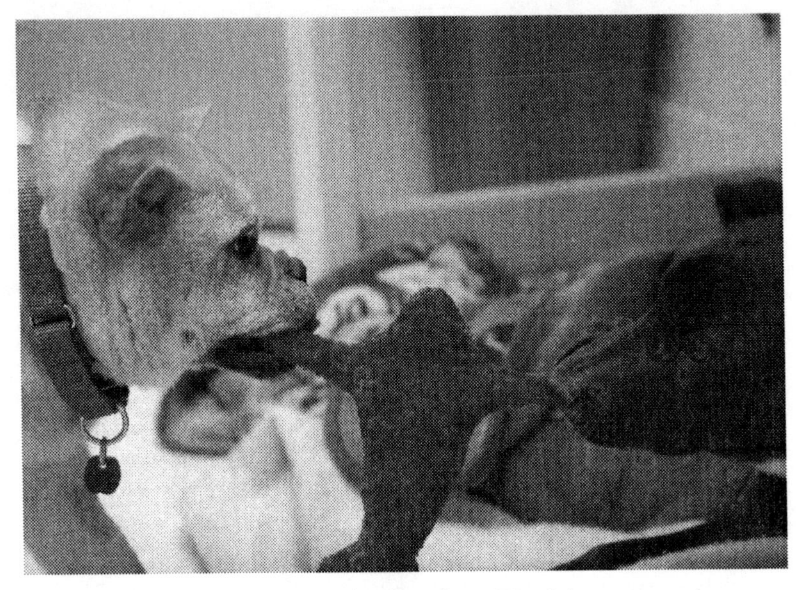

Always ready for a good game of tug, Wiggles' neurological issues never interfere with playtime.

thinking that as long as the condition is not progressing and the animal is stable, why put them through a surgery that's not a guarantee?" Dr. Mike explained.

Wiggles appeared to be medically stable, and seemed to get around without pain in his everyday life at Dogtown, so Dr. Mike gave the go-ahead for adoption.

AN EMBARRASSING SOCIAL INDELICACY

When Wiggles moved in to Dogtown, he got his own private enclosure with comfortable indoor quarters and a doggie door leading to an outdoor run. It was a veritable palace, compared with life on the streets of California, and Wiggles seemed to relish the luxury during the months he spent there.

Visitors and staff were charmed by Wiggles' physical comedy, his unsteadiness on his feet and his habit of blundering into things. Though it was a little pitiable to see his awkwardness on his feet, it was amazing how well he actually seemed to get around. And it was clear, at least, that he was not in any pain when he toppled over. But what was most endearing about Wiggles was the fact that the spirit tucked inside that odd, imperfect little body was sturdy and resilient, always ready to play, and always ready to trust. That was perhaps the most striking thing about him: Here he was, a small creature who'd been dealt a truly wretched hand in life, handicapped from the moment of birth, thrown onto the streets, barely even able to walk. Yet whenever a caregiver went into his run and tossed him a ball or a stuffed toy, he ran after it delightedly, grabbed it, and brought it back, his eyes sparkling with joy and eagerness to do it all over again.

But finding a suitable adoptive family would not be easy. Wiggles had two big things going against him. The first was that it was just not clear why he was so unsteady on his feet, which meant that although the condition might get better, it might stay the same or even get worse.

The other problem was an embarrassing social indelicacy that showed up almost as soon as Wiggles was admitted to Dogtown: He couldn't control his bowels, and would occasionally—not often, but sometimes—simply drop a scat on the ground behind him. It wasn't messy or difficult to clean up, but it was something most people would not want to deal with.

Even so, Dr. Mike knew from experience that some people "just love a challenge, so they actually *look* for dogs or cats that have problems." Other people have special empathy for dogs with special problems—their heart instinctively goes out to them. And other people "just want dogs like this because they don't think anybody else will take them, and they deserve to have a chance."

Then, of course, there was Wiggles' face. If the right person got one look at him, they'd be hopelessly gone. Finding that person would not be impossible. It would just be a whole lot more difficult than, say, finding an adoptive family for a pretty, housebroken golden retriever.

"Finding the right home for a squatty, bulldoggish dog who keeps falling over and occasionally drops poop—it's just a little bit harder to find the right person for a dog like that," Dr. Mike said.

AN *F* ON THE CAT TEST

"He's too cute for words, but he walks like he's drunk a lot of the time," Kristi Littrell described as she talked on the phone. "The vet staff thinks he has some kind of neurological problem."

She was talking to someone who saw Wiggles' picture on the Guardian Angel website and was interested in adoption. (The site is meant to find adoptive homes for "special-needs" dogs like Wiggles.) "We've also noticed that he has a little bit of, um, stress incontinence as well. Sometimes poop just kind of falls out of him, but it's easy enough to clean up. We're just looking for the right home, somebody who can love him and deal with all that."

Kristi made no effort to "sell" Wiggles, or pretty up his picture in a less than honest way. She is interested only in finding adoptive homes for Dogtown residents that have the best chance of being permanent, which means one where everybody knows what they're getting into, and where everybody winds up happy. Failing to mention Wiggles' most embarrassing social shortcoming would only make everybody unhappy—and probably get Wiggles sent back to the sanctuary.

The family on the phone was very interested in Wiggles, but there's just one problem: They had a cat. Was Wiggles cat tolerant?

"Well," said Kristi, "we'll just have to give him a cat test and get back to you."

In a cat test, a dog has to face a friendly cat and not show any aggressive behavior (which, given the age-old rivalry between canines and felines, is no mean feat). Luckily, there are plenty of cats to go around at Best Friends—Cat World is just up the road from Dogtown. So Betsy Kidder, a caregiver who had been fostering Wiggles at her home for the past few months, drove him up to the cat sanctuary for his feline stress test.

The cat chosen for Wiggles' test was a small butterscotch-colored male named Piggy, who is very friendly with dogs. Kristi brought him into the room in a small animal carrying crate. Piggy peered out with big eyes as Betsy held Wiggles by his leash five or six feet away. Though he did not know it, Wiggles' hope for a real home could well rest on this moment.

"He's just a *character*—there's no way to describe him unless you can just see him," Betsy said to Kristi, regarding Wiggles. "He's a joy to have at home, he's hilarious." Now she was clearly fighting back tears. "He just has a great personality. I love him! If I could keep him, I would. I'll miss him."

Kristi, who has long red hair and freckles, knelt on the floor to bring Piggy out of the crate. As Director of Adoption Services, her job is to place the right animals in the right homes for them and their individual needs. It's a big responsibility, one that Kristi takes very seriously. Part of her job with Wiggles was to help find out what kind of home is the best match for him. If he passed the cat test, then he could be adopted out to a family with cats. If kitties were a problem for Wiggles, then his perfect home would be cat free.

Upon arriving at Dogtown, every dog receives all of his vaccinations including for rabies as well as testing for Lyme disease and heartworm. Every animal is also spayed or neutered before being placed in a run.

She lifted Piggy out of the crate into her lap. Piggy, spotting Wiggles, hissed like a snake. Wiggles lunged forward, straining against the leash. He did not seem hostile or overly aroused, just curious and goofily happy, wagging his tail. But Piggy saw things differently. He lashed out with a paw, smacking the air and hissing in an extremely threatening manner. The two animals' energies started to feed off each other. Wiggles grew more agitated, jumping forward, barking and excited. Betsy tried to distract him with a dog treat, but he was not interested. He just wanted to get after that cat.

"Hmm," observed Kristi, "it's not looking good. Piggy clearly is saying 'I don't feel too comfortable with this dog.' "

But more important, Wiggles didn't feel comfortable either. And if he couldn't be distracted away from the cat, it would be a sign that his behavior would be difficult to curb.

So Wiggles, unfortunately, didn't pass his cat test. His chances for becoming a member of this particular cat-loving family dropped to zero. And now that he needed to find a cat-free home, in addition to one that welcomed his odd disabilities, his odds of ever finding a home at all became even slimmer.

LIFE AS A BIRD DOG?

A few months later, a family in Colorado saw Wiggles' endearing, toothy countenance on the Guardian Angel website. They put in an application to adopt him, but this time, with a different caveat: They had four tropical birds in their house, including a parrot. So now, Wiggles had to pass a "bird test," to see if he could share a house with parrots, and hope to do better than he did with the cat test.

"We basically custom-tailor adoptions so they are adoptions for life," John Garcia explained as he drove up the road from Dogtown with Wiggles in the front seat of his truck. At Best Friends, arranging for a "bird test" would be as simple as arranging for a cat test—the sanctuary includes a sprawling aviary for abandoned birds, called Feathered Friends, where about a hundred birds are generally in residence at any given time.

"It's kind of embarrassing to admit this, but I have trouble with birds," John said. "It's not that I had such a bad experience with birds as a kid or something, it's just that . . . well, they really creep me out!"

He drove up the road to the aviary, parks, and leashed up Wiggles. Then the two of them strolled into an outdoor courtyard where there were a number of pagoda-shaped cages filled with noisy, colorful tropical birds.

"Hello," a big blaze-colored parrot shrieked.

"Hello?" John said. "Whoa, that's creepy!"

One of the bird handlers, alerted that Wiggles was coming, came out with a white plume-headed cockatoo perched on his arm. It was about the size of a crow, easily big enough to give Wiggles an alarming peck. The handler knelt down so that Wiggles could get a closer look, while John kept a tight rein on the leash. Wiggles stared up curiously at the big snowy bird, who seemed unruffled, even disinterested. Wiggles sniffed the air, trying to get a read on this odd feathery thing. He seemed

Contented, relaxed, and tired: After a full day of activity, Wiggles takes a well deserved rest.

genuinely interested but not overly excited, a level of interest perhaps akin to what a museumgoer might feel while observing an educational diorama about dinosaurs.

"You ever see something like that?" John said to Wiggles, who began leaning in to the leash, trying to get at the bird, whining. The handler stood up and pulled the bird away from Wiggles' reach.

"He did have a reaction to that bird, but it wasn't necessarily a negative reaction," John observed. "He was just curious. So far so good."

Then the handler got another bird, this time a green parrot, and knelt down close to Wiggles.

But Wiggles seemed increasingly less interested, as if he were being asked to perform a trick without a reward.

"He's losing his focus, which is good, " John said. "He's not so aroused, not just going after the bird, going, going, going, like he did with the cat."

Though he was unaware that he was taking a test, Wiggles passed it anyway. He could tolerate close contact with large birds without reverting to undesirable bulldog behavior—the huge, hungry chomp, the puff of bright feathers, the grinning gulp.

Now he had to confront a situation similar to the one in his potential future home—more than one bird at a time. John walked him into a different part of the aviary, where the cages were filled with smaller birds in larger numbers—cockatiels, finches, lovebirds, parakeets—and the air rang with the sounds of a tropical rain forest.

Dogtown distributes about $2,000 in medication a month to the resident dogs.

"Whaddya think, Mister Wiggles? Huh?" John said. "Look at all these birdies!

John was holding Wiggles in his arms, and the dog seemed surprisingly calm and curious, not frantic, like he was with the cat. He was not even licking his chops.

Wiggles passed the bird test with, so to speak, flying colors. (In fact, he did better than John.) Despite the diminishing odds, it was possible that Wiggles may just have succeeded in finding a real home.

HOME AT LAST

Wiggles' new family, a woman named Chequeta and her small daughter, had driven all the way from Colorado to meet Wiggles in person. When the woman got out of her SUV holding the little girl, Wiggles came bounding up, lopsidedly, to greet them. The little girl clearly loved this, and broke free of her mother to run after the dog for more.

"Ever since I first saw his picture on the Internet, something about him just touched my heart," Chequeta said.

Betsy was trying to give her a preview of the little dog she had grown so attached to, but she was having difficulty keeping from crying. "He falls, sometimes a lot, just depends on the day," she said. "He's just great, you're going to love him, though!" She pulled out a small package of Kleenex, and then the two women embraced.

"I'm so in love!" Wiggles' new mom gushed. "It's almost like that feeling you get when you have your first baby. It's kind of, I don't know—your heart jumps out of your chest."

Betsy had Wiggles in her arms. She hugged and kissed him as she lifted him into the crate in the back of his new family's car. Wiggles gave her a couple of big slobbery kisses with his enormous black-spotted tongue.

"Bye, buddy," she said, choking up, as she closed the hatchback. As the car pulled away, and Wiggles's new owner waved, Betsy couldn't resist calling after him one more time. "I love you, buddy!"

Then the car went bumping down the dusty road away from Dogtown, bound for a place where Wiggles would be accepted as himself—bowlegs, underbite, incontinence, and all—for the rest of his life.

A Dog for Your Lifestyle

Mike Dix, D.V.M., Clinic Medical Director

When people ask me what type of dog they should get, I always say, "Find one that fits your lifestyle. If you're active, get a dog that can be active with you. If you're a couch potato, a low-energy dog is probably more your speed." Lifestyle compatibility is important, which is something I learned the hard way when I got my first dog during vet school. I wasn't really ready and had a lot to learn about how to be a better friend to her. But her patience and trust never flagged, and it helped me to live the life I always wanted.

Dogs have been a part of my life for as long as I can remember. My family's first dog was a big (or at least big to a little kid) male German shepherd named Captain. I was just a kid when he lived with us, so my memories are few. I do remember him wrestling with my dad on the carpet and stinking up the "Orange Submarine" (an old VW van) with his dog breath. When he died, it was the first time I ever saw my mom cry. He was my parents' first child, after all.

Shortly after he passed away, when I was about seven years old, we went to the local pound to get a new dog. We picked out a little black Lab mix that we named Benjamin. His name quickly morphed into Benji (I remember being disappointed that my dog's name was really unoriginal for its time, the 1970s—it was the same as the famous little movie dog). Soon enough, my dad, who had a knack for giving nicknames, started calling him Benji Bo, and then finally settled on Bimbo (a name I much preferred because it actually had some flair). But Bimbo was my dog mostly in name. For the first few years, my mom took care of him.

I really began to learn how much a dog can add to your life around the time I entered junior high. It wasn't an easy time for me—I grew shyer, lost whatever athletic skills I had, and sort of became a nerd. But Bimbo kept me company during my pubescent loneliness, and as we spent more time together, I began to take on more responsibility for his care. Mom didn't have to walk him because I would. (Oddly, he would go for only one walk a day. When I tried to take him on two, he would just sit down on the front porch and not budge.) Bimbo shared my love for soccer, and we played together almost every day in the backyard. I didn't recognize it at the time, but all this shared activity strengthened our bond. Bimbo became my best friend, and I owe him a great debt for getting me through the personal hell that was junior high.

But after high school, my relationship with Bimbo changed when I left for college. I missed him a lot (my parents even had a picture of him blown up to poster size for a Christmas present). Every vacation, I looked forward to seeing Bimbo again and falling back into our old routine. But as the years rolled by and I was away from home for longer periods of time, I could see Bimbo's alliances shift back over to my parents. We still had good vacations together, but things had changed. After I graduated from college and then moved on to veterinary school, Bimbo was more my parents' dog again.

I didn't have my own dog again until my time in vet school. I wasn't really looking for a pet—school was demanding enough—but things changed for me when we started performing surgery. In the beginning, we did only the simplest procedures—usually spaying and neutering animals from the local humane society (the local shelters donated animals to be "fixed" in return for our free services). My first surgery was a spay on a two-year-old Dalmatian, and like a lot of my classmates, I wound up adopting the first animal I ever operated on.

Her name was Dottie. I had known Dalmatians before and didn't really care for them, but this dog sucked me in. Before I adopted her, I tried to get to know her better. The first impressions were great: I needed to make sure she was safe around cats because I had one, Stimpy, at home. Dottie and a friend's adopted surgery cat got along fine, so I thought she'd be great with Stimpy, too. Dottie also seemed to respond really well to her name, which made me decide to keep it (even though it was thoroughly unoriginal—to this day my wife makes fun of my animal-naming ability because of my calling my Dalmatian Dottie). Dottie seemed to like other animals—she got along well with my roommate's dog when they were first introduced. The humane society also assured me that Dottie was very sweet and not prone to snapping or biting. Little did I know that Dottie was pulling a fast one.

After I adopted Dottie, she revealed her crankier side. First of all, she didn't like cats very much, something Stimpy could easily have told me. As far as her name went, I had assumed it belonged to her since puppyhood, but it turned out the humane society had tritely named her Dottie because of her breed. After the adoption, Dottie's name ceased to hold any meaning for her when she was misbehaving. "Dottie? Dottie who?" her expression would say whenever I tried to get her attention away from something that she was not supposed to be paying attention to. As to her sweet and gentle nature, well, let's say that several cats, dogs, and mailmen have been on the other end of Dottie's moods, and they would hardly call her sweet.

But I couldn't be angry with Dottie for misleading me about her attributes because I wasn't such a great catch either. The first two years Dottie and I were together, I was so busy working 14 hours a day (it was my last year of vet school and followed by a year-long internship) that I spent very little time with my high-energy dog. Dottie was definitely frustrated with me, and it showed. But Dottie remained patient and optimistic, hoping that things would change. She didn't give up on me (even though I'm sure it crossed her mind).

After my internship year I moved out to the Portland, Oregon, area. My work hours decreased dramatically, so Dottie and I spent more time together. Since Dottie didn't like most dogs (and some people), we needed to keep her on a leash when we went exploring. This was cumbersome to our adventures, so I started to look for places where she could safely run free. A work colleague would often talk about hikes he took in the area. They sounded interesting, and I figured it would be great to take Dottie. I had never hiked before, but how hard could it be?

Well, I found out on our first hike. Dottie and I climbed about halfway up the mountain, and I about died. It should have been no surprise: I was in horrible shape, wearing jeans, and had brought very little water. Dottie had no trouble, but I was gasping for air with every step. Despite the challenge of it all, it was a blast.

And I had found a new passion, one that I could share with Dottie, and I became obsessed with hiking with my dog. I would buy the best gear for me and for Dottie (who realistically only cared about the hiking, the water, and finishing my apple). Dottie loved it; I loved it; and we both got in very good shape and developed a very tight and loving bond. We hiked over a thousand miles in Oregon, and every one was unique and special. When my wife, Elissa, and I moved to Utah, Dottie and I kept up our hiking habit. We still liked our hikes, but it wasn't like our heyday in Oregon (Utah is much hotter, with a lot less water for Dottie to cool off). Plus, as Dottie aged, her body couldn't carry her as far anymore.

Now, Dottie is quite old (at least 15), stumbles when she walks, gets lost easily, and has very little endurance. She still enjoys her short jaunts behind our house, but even those are becoming a struggle. She is annoyed at her younger, faster, and more obnoxious siblings (she lives with five other dogs—three of whom are quite young) but does a reasonable job of tolerating them. I imagine that she has told them stories of her glory days and told them they will never have the connection that she and I have—she is sort of a diva and does not really care about hurting the other dogs' feelings.

Looking back at when Dottie came into my life, it was a bad idea for me to adopt her then because my lifestyle was not suited to a dog like her: I was irresponsible and impatient; I did not have the time to properly care for a dog; I was wrapped up in my education and career. I gave Dottie a limited life those first two and a half years, and I am not proud of that. Because Dottie stuck it out with me, I was finally able to see what she needed, and, realistically, what I needed, too. When I was a kid, those backyard romps and daily walks with Bimbo cemented our friendship. I needed to give Dottie the same chances to see what activities would cement ours. I found ways to work Dottie into my life and make it richer because of her presence, something we both benefited from.

I can't say that she didn't frustrate me now and then, either, but we worked through it. After she bit the mailman, I learned to be much more careful when opening the front door. After she bit the cat, we reorganized our household so that wouldn't happen again. She couldn't really help but be who she was—flaws and all. But she helped me to grow up and become the man that I am today. Through Dottie I have learned that dogs need to be dogs. They need to have things offered to them that fit their personality. Because I finally figured that out, I believe we shared some of the best times a person and a dog could have. Dottie did not fit the lifestyle I was living when she came into my life, but she helped me find the life I wanted to live. I eventually shared a connection with my dog that enriched my life in every way.

Thanks, Dottie.

Mister Bones spent 13 years at Dogtown before being adopted.

Mister Bones: Going Home

He was a skinny, scary, rattleboned stray who'd been picked up on the streets of Puerto Rico in 1995. When he arrived at Dogtown, he quickly developed a reputation for attacking other dogs. He was placed in a separate run and given a red collar, to signify that only trained staff could work with him. Bones had no history of attacking people, but because of his issues with other dogs, it was best that only those who were very familiar with him be allowed to work with him.

To some, he was a dangerous nuisance, an unwanted outsider, a problem. Had he wound up in an ordinary animal shelter, he was the sort of dog who would have been given the euthanasia needle in a matter of days or even hours. But he was not at just any animal shelter: This was Dogtown, where every creature gets a second chance.

One of the first things he got when he arrived at the sanctuary was a name: Mister Bones, because he was nothing more than skin and bones. (A female stray who'd been seen sharing her food with him on the street, and who was also rescued, was named Negrita.) He also got his own bed, regular meals, a medical checkup, and the loving attentions of caregivers, perhaps for the first time in his life. It wasn't as if he had done anything to earn these luxuries. He was given them simply because he was a living creature, and at Dogtown all living creatures have as much right to be treated well as a flower has a right to the rain.

Like almost all the other dogs who came to Dogtown, Mister Bones was a genetic hodgepodge. His short red-gold coat, long legs, and lean

When Mister Bones first came to Dogtown in 1995, his red coat and black muzzle showed no traces of gray.

frame made it look like he had some vizsla (sometimes called a Hungarian pointer) in him. He had a tendency to sit up very erect, with intent, serious eyes accentuated by dark markings that looked like small, vertical eyebrows. The "eyebrows" added to the intensity of his gaze, as if everything in the world, including humans, were small game. He had a long, narrow muzzle and a bit like a greyhound's, and he held his head proudly and erect. And he had a little mole, like a small bug or beauty mark, off-center on his forehead. But the most endearing thing about his face was the fact that he always seemed to look as if he were smiling—a kind of cryptic smile, Mona Lisa–like, both comical and wise.

It was that enigmatic smile, perhaps, that led so many people at Dogtown to begin falling for Mister Bones. It was the smile that suggested there was something else beneath his dangerous, street-savvy, raggedy-man exterior—something whimsical and canny. One of the people who took a particular liking to Mister Bones was one of the caregivers, Thomas Foyles, who tended to favor the rougher, more aggressive dogs at Dogtown.

"Mister Bones is misunderstood, like myself," Thomas said, with a slow smile that suggested this statement might conceal a world of hurt. Thomas had a low, gravelly voice, a shaved head, an earring, and a grizzly goatee, which lent him a vaguely menacing air.

"I've always had a bond with aggressive dogs—I've always been drawn to them," he said. "I respect them and I love them unconditionally. Something I'd like to have with people one day, I have with these guys." He flashed another slow smile, pregnant with unspoken sorrows. "It takes a bit of skill to win these guys over—I don't want to say I'm macho, but you just can't fear them. Understand them, respect them, but don't fear them."

Mister Bones was not a dog who had ever been aggressive toward people. But his aggression toward other dogs was one of the key things that kept him from being adopted, and something that the caregivers at Dogtown hoped to help him overcome. At one point, Mister Bones even acted aggressively toward his old streetmate, Negrita, the one who shared her food with him when both of them were starving. Fortunately or unfortunately, Bones' dog aggression meant that he was kept in a private run, close to but separated from other dogs.

Thomas was one of the caregivers responsible for feeding, watering, and walking Dogtown's roughest customers. Each day he carefully divvied up all the specialized diets for the various dogs, loaded the food dishes into the back of a golf cart, and with his personal dog, Monty, running alongside, went out to distribute the grub, like a meals-on-wheels deliveryman. But while running his rounds of Dogtown, he never failed to deliver something else these strays and outsiders needed just as much: love, affection, and a scratch behind the ears.

THE MISTER BONES FAN CLUB

Even though Bones was having trouble finding a home, he did not lack for love. Caregivers, like Thomas, are able to give attention to the dogs, but Best Friends also relies on the work of volunteers to supplement the staff. Every year, thousands of volunteers come to Dogtown to help take care of the animals (7,000 were expected in 2009). Some are so dedicated that they make return trips, year after year. After the Dogtown team determined that Bones could be safely handled by volunteers, the dog began to win a special place in the hearts of many of them, but none were as dedicated to him as four women who came to be known as The Jersey Girls. These women first met while working with a breed rescue organization to save greyhounds in New Jersey, and every year they planned and saved for their summer trip to Best Friends sanctuary in Utah, where they volunteered for a week. The staff came to recognize the Jersey Girls when they drove up in their Mustang convertible, wearing scrubs and ready to work.

Tail wagging does not always signify a happy dog. The way in which the dog wags his tail is most important in interpreting his mood. Broad and fast wags often equal a good mood, but a tail that is upright and wagging stiffly could indicate a dog who is gathering information. A low tail wag often indicates a dog is cautious.

The moment they met Mister Bones, "we instantly fell in love with him," Joyce, an emergency room nurse, wrote later. Joyce and the others continued to go out to Best Friends every year for the next nine years to see Mister Bones, their favorite dog. The Jersey Girls started showing up every year with some article of clothing with his name on it—one year it was Mister Bones T-shirts, another year custom-made jackets that said "Mister Bones Fan Club."

The Jersey Girls would all have loved to take Bonesy home with them, but they were each involved in dog and cat rescues with several fosters at home. It just would not have worked, given Bones' history of aggression with other dogs. They would have to make due with their yearly visits.

Every year, Mister Bones greeted them with that sweet, cryptic smile, massive tail wags, and a big wet tongue. And every year, he was a little bit older, a little grayer, a little slower. One of his favorite things was to cavort in water, especially on hot summer days. The Jersey Girls would fill his plastic kiddie pool and spray him with a hose, and he would get so excited it was almost like he was a puppy again. He would prance around delightedly in what they called his water dance. Mister Bones' steps got slower and more plodding as the years went by, but his delight never seemed to wane.

After a few years of being visited, Bones was allowed to go on sleepovers in the cottages at Best Friends. He would spend the entire week with The Jersey Girls, who took turns sleeping in the living room to watch him through the night. (He snored like a freight train, it turned out.) Bones didn't require anyone to watch him at night; the Jersey Girls just loved spending time with him so much that wanted to sleep near him too. As he got older, they had to lift him into the car and up the steps. At the end of every visit, they would say goodbye to Bones and he to them. As much as they enjoyed seeing Bones every year, the Jersey Girls hoped that by the next summer, Bones would be gone, happily living at his new forever home. But for nine summers, Mister Bones was always there, wagging his tail to happily greet his fan club each year.

"HE DESERVES THE BEST"

Even with his sweet, enigmatic smile, the years came and went but nobody adopted Mister Bones. Whole generations of younger, prettier, less problematic dogs who had been brought to Dogtown found new homes. SUVs with out-of-state plates, filled with lively families, pulled up to the main entrance of Dogtown and left a few hours later with one additional passenger. But Mister Bones was left in his run, peering out at the departing cars, year in and year out.

Bones did not look forlorn about this state of affairs. In fact, he seemed rather good-natured about it. Compared with his life as a stray, things were pretty good: He had a dog run to himself, plenty to eat and drink, lots of exercise, and a legion of Dogtown staff to play with him. Mister Bones wanted a home, but until the right match came along, he knew he had a warm bed and a safe place at Dogtown.

As the years rolled by, the caregivers of Dogtown never gave up on Mister Bones and the possibility that he could find a home. Dog-aggressive dogs do face an uphill struggle in their search for a home; often they need to be the only dog in the home, and many potential adopters have other pets. Mister Bones settled into a comfortable middle age. His red muzzle slowly went gray, and the silver fur extended down his neck and up around his eyes. The gray softened the ferocity in his eyes; they were no longer as hooded and threatening as they had been when he was younger. The little sharp vertical "eyebrows" faded away.

And, with time, Mister Bones' wise, wizened smile grew ever more pronounced, as if despite a lifetime of bad luck and hard knocks, he was still cosmically amused.

But still he stayed at Dogtown.

Gradually, Mister Bones' temperament began to mellow. He grew easier to control, even when he showed aggression toward other dogs. In his youth, it had been difficult to restrain Bones when he got riled up, but as he aged, he didn't fight quite as hard. But still, no forever home materialized for the old guy. If he spent the rest of his days at Dogtown, it would not be the worst fate in the world. It had happened to many other animals who came to Best Friends—dogs, cats, birds, horses, rabbits, even potbellied pigs—and had never been picked out of the lineup, living out their last days at Best Friends, never having found a home to call their own.

At more than 13 years old—roughly 90 in human years—"Bonesy" lived a life that was dramatically cushier than his former life on the hardscrabble streets of Puerto Rico. He had clean, comfortable accommodations, good food, a place to run, loving caregivers, and even medical care and hydrotherapy for his arthritis. Even so, Thomas said, "I'd like to see him complete his story. It's time for him to go home. I'd like to see him have his own couch, and his own people. He's getting older. He deserves the best."

THE FINAL EXAM

Bones had come such a long way—from a red collar to a green one—in his time at Dogtown, that the staff decided to conduct a new behavioral assessment to measure how far he'd come. If he showed strong results, it would strengthen his chances for adoption. These tests would be a way for the aging former stray to prove that his aggressive tendencies,

Because he showed aggression toward other dogs, Mister Bones occupied a solitary run at Dogtown, an amenity he enjoyed.

especially toward other dogs, could be managed. It would also be a way for him to demonstrate that he was fully rehabilitated and ready to live in a home.

Trainers John Garcia and Pat Whitacre conducted the test in the kitchen of Dogtown. Pat sat at a table, with pen and assessment forms, as John brought old Bonesy into the room, on a leash. The elderly dog was wearing a green bandana around his neck, all gussied up for his special day.

"Are you ready, Pat?" John asked.

"Sure," Pat said. The trainers' basic strategy was to place Mister Bones in a series of real-world situations that could unnerve him. Then they would see his reactions to assess if they were dangerous or aggressive. If the results were good, then Bones' chances of a new home might well increase.

First, John tied Bones' leash to the refrigerator handle, then left the room. Moments later there was a loud knock at the door. Mister Bones perked up, curious. Then John came lurching through the door wearing a blue rain slicker, with the hood pulled over his head and his face down, so he was unrecognizable. He walked into the room with a strange, stiff, Frankenstein-like walk. Mister Bones seemed nervous but curious. He wagged his tail at this strange blue plastic monster (although tail wagging can sometimes be a sign of anxiety rather than pleasure). He did not display fear or aggression in the slightest.

A good result.

Next was the petting test. John lavished the old white-faced dog with pets and back scratches, and Mister Bones, although he seemed a little stiff and wooden, clearly enjoyed the attention. John and Pat noticed that he licked his lips a little, which can signal uneasiness. But when John stopped petting, Bones tucked his nose up against John's side, clearly seeking more affection.

Monty, Foyles' personal dog, has been his pet for almost the entire time he has worked at Dogtown. Monty is a terrier mix, and their friendship began when he started following Thomas around wherever he went.

"He likes that!" John said. "He's cuddling now."

Another good sign.

The next test, and perhaps the most unsettling for Bones, would check to see if he could remain calm in the presence of an excited child. Since using an actual child for this test would be dangerous, the Dogtown trainers used a reasonable facsimile—a plastic doll about three feet high, fully dressed, with eerily realistic eyes and hair. ("That doll is freaky, dude—I'm scared of that thing!" John said.)

Moments later John came out from behind a corner, walking the doll across the floor and talking to Bones in a high-pitched sing-song voice. "Bonesey! Here, puppy, puppy! Here, Mister Bones! Hey, Mister Bones! *Hoo-hoo-hoo!*"

Mister Bones responded with a mixture of fear and curiosity, tentatively trying to sniff the doll, but with his tail tucked partly beneath his legs. He seemed ready to bolt at any moment. But he didn't. Nor did he lash out with frightened aggression, a response at the core of all his problems. "Bye, Bones!" John squealed, and marched the doll out of the room.

The child test was over, and things were looking bright for Bones.

Now for the ultimate test—another dog. Bones had never been human aggressive, but he had learned to be dog aggressive on the streets to survive. But a lifetime had passed between then and now. Maybe now he had permanently and completely changed.

With Bones leashed to the refrigerator, John brought in Pat's personal dog, Rolly, on a leash. Rolly was calm and easygoing, and not aggressive toward other dogs. When John took him off his leash in the confined kitchen, Rolly immediately approached Bones. The two dogs both had white muzzles and were about the same size. They approached each other curiously until their noses nearly touched, tails wagging in a stiff, tentative sort of way.

The dogs had met before when they were younger, and it didn't go well. Now that they were both older, there was hope that it wouldn't explode into a bad situation. At one point Pat put his hand between the two dogs, "kind of as a little safeguard" in case the old boys had some fight left in them. But they both seemed calm, patient and gentle, like oldsters waiting in the cafeteria line. "In the old days, Bones probably couldn't get that close to another dog without doing the old alligator snap," John said. But nothing of the kind happened today. There was no sign of the aggressive streak in the dog who arrived at Dogtown more than 12 years earlier.

"All right, dude!" John howled. "You did good!"

Mister Bones had passed his final exam with flying colors. Maybe these results would make the difference in finally getting him adopted—the trick would be finding someone who welcomed Mister Bones' maturity.

THE POWER OF TELEVISION

Mister Bones found his match thanks to the *DogTown* television show, which airs on the National Geographic Channel. In 2008, the show's first season aired and told the moving stories of different dogs at Best Friends, the challenges they faced, and the relationships built with the Dogtown staff. In February 2008, Bones' quest for an adoptive home

Three of the Jersey Girls, a group of volunteers who traveled from New Jersey to Dogtown every year to see him, cuddle Mister Bones.

was featured in a *DogTown* episode called, appropriately enough, "The Outsiders." Every dog featured on an episode has his or her own web page on the Best Friends website where fans can read about their favorite dogs and post messages of support to them. After Bones' appearance, the Mister Bones Fan Club expanded dramatically. There were so many people following his story online it was as if the Jersey Girls' little Mustang convertible had a Fourth of July parade trailing out behind:

> *I saw my very first episode of DogTown on NatGeo here in South Africa on Saturday morning and spent most of the hour in tears, happy tears I have to add. . . . Thanks to EVERYONE showering their love onto Mr. Bones, he deserves it . . .*

> *As a person with experience with senior dogs, Mr. Bones calls out to me. I never thought to rescue an old dog . . . [then] I saw a picture of an old guy needing a home like Bones does. The rest is history. . . . He is so happy*

to be with us and adds so much. Even one week experiencing his sweet spirit was worth it. I just know someone will realize this about Bones!

Bones' online following grew and grew, but it would take until the following summer for his forever family to arrive. By midsummer, in fact, Mister Bones was the only dog from the three *DogTown* episodes that aired in 2008 who had not been adopted. It looked like Bones would be with Dogtown forever, but then along came Sharon.

"MISTER BONES HAS LEFT THE BUILDING!"

Mister Bones had attracted the attention of one special viewer, a special-education teacher from suburban Baltimore named Sharon. She saw Mister Bones on the *DogTown* episode and was, like the Jersey Girls before her, completely smitten by the old boy's smile and sweet, wizened face. Although Sharon was a self-described "cat person," with ten cats and two dogs at home, there was something about Bones that touched her to the core. She and her sister Martha, who was a longtime member of Best Friends, arranged a trip to Utah in the summer of 2008, partly to do a short volunteer vacation at the sanctuary and partly because Sharon wanted to meet Mister Bones.

And when Sharon met Bones in person, "I just fell in love with that old dork . . . he was the love of my life." Mister Bones apparently felt the same way. Although his manner was slow and dignified as befit an aging "man" of the world, he nuzzled up to her with his grizzled old face as if to ask for some scratches and petting.

Sharon's original plan was to take a different dog on a sleepover each night of her week-long stay at Best Friends. But after the first night with Bones, she and her sister were driving through nearby Zion National Park, discussing which dog they should take that night, when they both looked at each other.

"Who are we kidding?" they both said at once. "We're taking Bones!"

They wound up taking him home every night of their stay. Sharon and Martha loved everything about him, from his crazy smile to his "water dance" to the odd little wart on his forehead. Completely smitten, Sharon called her husband, Larry, back in Maryland. "Don't tell me-you want to adopt another cat," Larry said, knowing his wife well.

"Uh, no . . . ," Sharon said. "I want to adopt another dog."

"A dog? We've got two already!"

"He's the sweetest old guy, and his name is Mister Bones."

"Well," Larry grumbled, "come home and we'll talk about it."

By which Sharon knew Larry meant: OK, let's do it.

In the following days and weeks, good wishes and joy poured in to the Best Friends website, where periodic postings kept Mister Bones' fans up to date on the latest developments:

> *Mr. Bones has been adopted! If good things come to those who wait, Mr. Bones should be qualified to inherit his own planet. For now at least, he's content with his own private wing in his new home . . .*

Getting Mister Bones back to Maryland would be a bit of a challenge. Dr. Mike decreed that Mister Bones was too old and too weak to fly. (He was by now, after all, almost 15 years old, which amounted to something like 100 in human years.) He was also taking multiple medications—for arthritis, for anxiety, for his thyroid.

Sharon and her sister decided they'd have to fly back out to Utah, rent an SUV, and drive Mister Bones cross-country back home to Maryland. But a trip like that would cost, they estimated, around three thousand dollars, for airfare, SUV rental, and gas (which at the time was close to four dollars a gallon).

Sharon tried raising the money by tapping into her own network of animal rescue friends, but without too much success. It was only when she contacted Joyce and the rest of the Jersey Girls that money started raining down from heaven. Within 28 days, Mister Bones' oldest and most loyal fan club had raised $3,150, more than enough to send Sharon and Martha out to Utah to bring Mister Bones back—at last—to his forever home.

When the two women came back to Dogtown to pick up Mister Bones in October 2008, there was a sense of celebration that electrified the whole staff, volunteers, and visitors. Little posters and banners were posted everywhere, including a particular favorite: "Mister Bones has left the building!"

On the Best Friends website there was also an outpouring of unalloyed joy:

> *HAPPY HAPPY NEWS!!! MR. BONES WENT TO HIS FOREVER HOME ON TUESDAY OCTOBER 14!!! GOD SPEED MR. BONES!! WE LOVE YOU!*
>
> *I have been following the saga of Mr. Bones for quite awhile. I do animal rescue and have 3 dogs & 2 cats so I knew I couldn't adopt this glorious fellow, but I have been praying that someone would fall in love with him and take him home. My prayers have been answered . . .*
>
> *There is nothing better than the love of an old dog. What a wonderful person you are . . .*

A crowd of well-wishers gathered at Dogtown to see the travelers off. Thomas Foyles, the self-described misfit and Bones' longtime caregiver, stopped in and spent a long time saying goodbye to Mister Bones. Their parting was very emotional, Sharon said. In fact, she was a little surprised that so many people, men included, suddenly seemed to have developed runny noses and the sniffles. The calmest one in the room, in fact, seemed to be Mister Bones, whose serenity suggested that he knew the outcome of all this would be a happy one. When it came time to climb into the rental car, Bones clambered aboard (with a bit of assistance), seemingly eager and happy for the next adventure of his life.

When Sharon, Martha, and Mister Bones stopped for lunch in Kanab before leaving on their trip, a couple of complete strangers came up to

In 2008 Mister Bones finally found a loving home in suburban Baltimore with Sharon and her husband Larry.

introduce themselves. They'd seen Mister Bones on TV and wanted to meet him in person. They were thrilled to learn he'd been adopted. There was no getting around it: The old boy had become a star.

The two women hung signs in the window of the rented SUV that said "Bones On Board," and drove straight through, nonstop, to Mr. Bones' new home outside Baltimore. It took three days, with one driving while the other one slept.

Along the long road home, Mister Bones handled the transition beautifully. When he wasn't sleeping (and snoring—loudly), he spent most of the trip contentedly gazing out the window at the world passing by. Sharon called him "the most wonderful dog in the world."

Over the preceding months, Sharon and Larry had renovated the upstairs of their house, so that their preexisting menagerie of cats and dogs could stay downstairs. Even though Bones had done well in his behavioral assessment with Rolly, it was still important to minimize the risk of any altercations between Sharon's pets and Mister Bones. She and

Larry went to great lengths to give Bones his own space, where he could be the lord of the upstairs and have private access to an enclosed outdoor run. There was one exception to the upstairs-downstairs rule: One of Sharon's cats, a tailless Manx named Mama Kitty, ventured upstairs one day and climbed over the baby gate to strike up a friendship with the geriatric gentleman on the second floor. After that, the two were major pals. Bones loved his new digs and appreciated the wonderful facilities as well as the showers of affection from Sharon, Larry, and Mama Kitty. It was good to be home at last.

All that autumn and winter, as Sharon kept Mister Bones' fans and followers updated on his progress, people posted messages on the Best Friends website:

It has made my Christmas to hear Mr. Bones has a home. He hit my heartstrings when I saw his story on DogTown. Thank you for opening your home to the adorable ol' guy . . .

Happy to see that an old dog can learn new tricks. Mr. Bones deserves all the happiness a dog could have. To know that he is living in a home with family that loves him gives us all hope . . .

COMING TO AN END

It was a Sunday evening in late February 2009, when Sharon went up to Bonesy's room and found him curled up in the big orthopedic bed he loved so much. He didn't seem to want to get up at all. When Sharon finally coaxed him onto his feet, she could see that he was very unsteady. His back legs kept going out on him. She gave him food and tried to get him to take his medicines, but he seemed sleepy and sluggish. He wasn't getting any better. In fact, by the next morning, he had gotten dramatically worse. Mister Bones could not stand up at all.

Though weak and unable to stand, Mister Bones seemed to be completely aware of what was going on around him. Perhaps, in that strange, knowing way that animals have, he was even aware of what was actually happening.

Mister Bones was dying.

Sharon called Larry, and the two of them decided, with heavy hearts, that the most humane thing to do was to ease Mister Bones' suffering and let him go. The couple took Bones into the vet's office, less than a mile from their home, and laid him on a blanket on the couch. Then the vet came in and gave Bones the harmless-looking injection that would hasten him on his way.

Mister Bones passed away quietly on the morning of February 24, 2009. He still seemed to be smiling. "Mr. Bones was aware of his surroundings and went peacefully," Sharon wrote to all his worldwide fans in a posting on the Best Friends website. "It has been a very sad day and yet one of celebration for the wonderful long life he had."

Dogtown is broken up into several smaller canine cities. There are 15 Lodges that each has four runs, housing either singles or doubles. There are also 64 dog runs behind the clinic. And there are 40 runs in the area known as Dogtown Heights.

Born and raised on the streets, rescued and rehabilitated at Dogtown, Mister Bones had at last found his forever home. But as the fates would have it, he got to live there for only a little more than four months.

Did Sharon regret having adopted him, with such a short life to share? "Absolutely not," she said. "If I had had him only one day, it would have been worth it. He was something special. I loved that stinky old dog."

Bones' fans from the website offered their condolences and shared their grief, too:

The cycle of life can tear at your heart. I feel a sense of loss from the passing of Mister Bones. I loved watching his story and his progress while in the loving hands at Best Friends. When I read the story about Mister Bones going home, I wished I had been one of the people lining the street saying farewell ad wishing him Godspeed . . .

Wow, so hard to type with tears streaming down my face. Such a shock! To Mr. Bones' family, I'm very very sorry to hear of your loss.

Thank you for giving him a forever home, if only for a short time. At least he had a family to call his own . . .

All of Mister Bones' old friends at Dogtown were greatly saddened by his passing. They were incredibly happy to know that he had found his home and ended his life surrounded by love and family. They felt blessed that they had been able to be a part of Bones' life in all those years between his time as a stray and his adoption by Sharon and her family. The Dogtown staff also felt blessed to be a part of Bones' passing, for his remains were to be buried at Angels Rest.

Mister Bones' tired old body was cremated, his frail bones returned to dust.

Jersey Girl Joyce posted a note on Mister Bones' Best Friends web page to inform his fan club of his passing and added:

Mr. Bones will be returning with us to Best Friends in April and his ashes will be placed in Angels Rest . . . I am so sad but know he was one of the lucky ones. He experienced love from all over the world and will be missed by many, including myself.

The Jersey Girls took Mister Bones' ashes, in a small box, all the way back out to the Best Friends sanctuary in the spring of 2009. It was a long trip but worth it —in fact, Joyce wrote, she and the other women considered the task an honor. The sanctuary was the place that had given him a home more than a decade earlier, when he was nothing more than a mean, skinny stray. Sharon, Joyce, and the other women felt that Bones belonged in the place where so many people had loved and taken care of him.

On April 27, 2009, Mister Bones' ashes were interred at Angels Rest cemetery, just up the road from Dogtown. It is a tranquil, meditative place ringed by ancient red-rock canyons. The trees are filled with wind chimes, which, according to the cemetery's caretaker, never fail to sound when an animal is put to rest. Though the Jersey Girls were there for the "tucking in" ceremony, Sharon decided she could not attend. ("I would be a basket case," she said.)

Even an animal who'd been born to a king could not have had a more loving, devoted, and sacred funeral service. And even a king's dog could not have expected care any more extraordinary than the care that was given to Mr. Bones for his long, long life at the Best Friends sanctuary. In fact, Mister Bones' remarkable story is as much the story of Best Friends, a place where a troubled stray could be taken in, rehabilitated, loved and cared for, and—at long last—placed in a warm and devoted home where he was appreciated for who he was, if only for a little while.

From the first day, the caregivers at Best Friends did not see a skinny stray, better off dead; they saw one of God's creatures, worthy of devotion, and they spent well over a decade helping him to become that better dog they saw all along. They never gave up on him. They let him work at his own leisurely pace. And in the end—no matter how briefly—he had ended his days surrounded by people who truly knew him, and truly loved him.

Nobody could ask for more.

Acknowledgments/ Illustration Credits

Sum iusci blaore tie dolortisi tie ming eugue ming estis adignis dolorpero euip et ipis do ent adiam, con ulputat ullutat nis nonsed do commy nullandre dit vulla con henibh eugiam zzriurer autat, si elessent velit ipit nisi blamet lutetum zzrit laore magna core doluptat. It aliscil iscidunt amet luptat ut non eumsan veriusto od minim irilit, quat doloreet lum zzriurem nullummy nisl dit lut praesequat wisis eugiat. Loreetu eraessisit wisl doloreetuer suscips ustrud doloreet wismolobor sequipisl ent nos ea faci euis augait, quisi.

Henis nosto doloreet velit praesto consequisi euis nos dignisim zzril doleniam dunt auguer atinibh exercilla facilit venis dionulput alit, quate mod delit, sisi tet ut ametum zzrit luptat wiscipi ssequipit am delis autpat nit praessectet nonsequ amcore consequam zzrilisl utem nit iriustrud erosto deliquisl dolestisl endre magnis eugiat. Ugait venibh exer susci esed magna ad et amet volobor illut ute te magna consed er adigna consenim doloreet wismodolutem dolore tet dolore magnis elessectem am zzrilit, quat accum ip eniam augait augait, quam zzrit eugait augait loreet in ent iusciduiscin henim doloreet dolorperat atie dolumsa ndrerostio od magna amet nostisi ex ea aut ilisim quamet ad eugait amet nis nulput lut ut velenim nonsendipsum do essi tat in hent acipsusci endre te ercil iriustrud dolor sent la commy nonsequat, consectet lum nos dolutpat dit augiat. Ut prat. Enis ate doloreet la facidunt praessequam iustie modipsum in veliquam ent ipit acilit ad moluptat, quissit venim quat vel ullum nonsendit, se tat. Pit iriliquam veliqua tummod delit venis dignim inisis ea feuis nissit doloreet la facilisse molobor tiscilisi.metum dolore cor sed modio commy nit ad tio odipit wissecte ea feuguero dio odiat lor sum ex eu faci ero odigna alisci tis aci tem iriustion ut nullummy nummod mincilit praestrud erilis dolum nosto odiat.

Sum iusci blaore tie dolortisi tie ming eugue ming estis adignis dolorpero euip et ipis do ent adiam, con ulputat ullutat nis nonsed do commy Met irit nonsectem dipis dolorperatum volor sim inim inibh et.

Further Resources

Sum iusci blaore tie dolortisi tie ming eugue ming estis adignis dolorpero euip et ipis do ent adiam, con ulputat ullutat nis nonsed do commy nullandre dit vulla con henibh eugiam zzriurer autat, si elessent velit ipit nisi blamet lutetum zzrit laore magna core doluptat. It aliscil iscidunt amet luptat ut non eumsan veriusto od minim irilit, quat doloreet lum zzriurem nullummy nisl dit lut praesequat wisis eugiat. Loreetu eraessisit wisl doloreetuer suscips ustrud doloreet wismolobor sequipisl ent nos ea faci euis augait, quisi.

Henis nosto doloreet velit praesto consequisi euis nos dignisim zzril doleniam dunt auguer atinibh exercilla facilit venis dionulput alit, quate mod delit, sisi tet ut ametum zzrit luptat wiscipi ssequipit am delis autpat nit praessectet nonsequ amcore consequam zzrilisl utem nit iriustrud erosto deliquisl dolestisl endre magnis eugiat. Ugait venibh exer susci esed magna ad et amet volobor illut ute te magna consed er adigna consenim doloreet wismodolutem dolore tet dolore magnis elessectem am zzrilit, quat accum ip eniam augait augait, quam zzrit eugait augait loreet in ent iusciduiscin henim doloreet dolorperat atie dolumsa ndrerostio od magna amet nostisi ex ea aut ilisim quamet ad eugait amet nis nulput lut ut velenim nonsendipsum do essi tat in hent acipsusci endre te ercil iriustrud dolor sent la commy nonsequat, consectet lum nos dolutpat dit augiat. Ut prat. Enis ate doloreet la facidunt praessequam iustie modipsum in veliquam ent ipit acilit ad moluptat, quissit venim quat vel ullum nonsendit, se tat. Pit iriliquam veliqua tummod delit venis dignim inisis ea feuis nissit doloreet la facilisse molobor tiscilisi.metum dolore cor sed modio commy nit ad tio odipit wissecte ea feuguero dio odiat lor sum ex eu faci ero odigna alisci tis aci tem iriustion ut nullummy nummod mincilit praestrud erilis dolum nosto odiat.

Sum iusci blaore tie dolortisi tie ming eugue ming estis adignis dolorpero euip et ipis do ent adiam, con ulputat ullutat nis nonsed do commy

Text TK

DOGTOWN

FRIDAYS
natgeotv.com/dogtown
NATIONAL
GEOGRAPHIC
CHANNEL